Distortion Theorems in Relation to Linear Integral Operators

Mathematics and Its Applications

Volume 385

Distortion Theorems in Relation to Linear Integral Operators

by

Yûsaku Komatu
formerly of Tokyo Institute of Technology, Tokyo, Japan

KLUWER ACADEMIC PUBLISHERS
DORDRECHT / BOSTON / LONDON

A C.I.P. Catalogue record for this book is available from the Library of Congress.

ISBN 0-7923-4304-2

Published by Kluwer Academic Publishers,
P.O. Box 17, 3300 AA Dordrecht, The Netherlands.

Kluwer Academic Publishers incorporates
the publishing programmes of
D. Reidel, Martinus Nijhoff, Dr W. Junk and MTP Press.

Sold and distributed in the U.S.A. and Canada
by Kluwer Academic Publishers,
101 Philip Drive, Norwell, MA 02061, U.S.A.

In all other countries, sold and distributed
by Kluwer Academic Publishers Group,
P.O. Box 322, 3300 AH Dordrecht, The Netherlands.

Printed on acid-free paper

Printed in the Netherlands

Table of Contents

Preface

The present monograph consists of two parts. Before Part I, a chapter of introduction is supplemented, where an overview of the whole volume is given for reader's convenience.

The former part is devoted mainly to expose linear integral operators introduced by the author. Several properties of the operators are established, and specializations as well as generalizations are attempted variously in order to make use them in the latter part.

As compared with the former part, the latter part is devoted mainly to develop several kinds of distortions under actions of integral operators for various familiar functionals; absolute modulus, real part, range, length and area, angular derivative, etc. Besides them, distortions on the class of univalent functions and its subclasses, Carathéodory class as well as distortions by a differential operator are dealt with. Related differential operators play also active roles. Many illustrative examples will be inserted in order to help understanding of the general statements.

The basic materials in this monograph are taken from a series of researches performed by the author himself chiefly in the past two decades. While the themes of the papers published hitherto are necessarily not arranged chronologically and systematically, the author makes here an effort to arrange them as orderly as possible. In attaching the importance of the self-containedness to the book, some of unfamiliar subjects will also be inserted and, moreover, be wholly accompanied by their respective proofs, though unrelated they may be.

While the materials dealt with in the latter part are fairly classical, the formulation in the former part seems new when it is put in the context of the latter part. The author expects that our framework will be supplemented by new examples giving further developments in near future.

In the preparation of this monograph, my sons Yū and Gen have greatly assisted the author. Full facilities for the publication have been given by Dr. Paul Roos, Acquisition Editor in Mathematics Division of Kluwer Academic Publishers, and his Assistants Ms. Angelique Hempel and Ms. Anneke Pot. To all of them the author expresses his deep gratitude.

May 1996

The author

Chapter 0. Introduction

§ 1. Overview

An analytic proof of a theorem is done in general by giving estimates, and to get sharp estimates almost always clarifies nature of the problem being studied. A "distortion" theorem is that of determining precisely the range of a real-valued functional $\Omega[f]$, such as $\max|f(z)|$, $\max|f'(z)|$, and so on, defined on a class of holomorphic functions $f(z)$, where we restrict ourselves to function s $f(z)$ in the unit disk $E = \{z \mid |z| < 1\}$. A typical class is $\mathcal{F}$ which consists of holomorphic functions $f(z)$ in E normalized by $f(0) = 0$ and $f'(0) = 1$. The following subclass is more familiar:

$$\mathcal{S} = \{f \in \mathcal{F} \mid f \text{ is univalent in } E\}.$$

The term "distortion" comes from a geometric interpretation of $|f'(z)|$ or $|f'(z)|^2$ as the infinitesimal magnification factor of arclength or area under a mapping f. The Koebe distortion theorem states that if $f \in \mathcal{S}$ then

$$\frac{|z|}{1+|z|^2} \leqq |f(z)| \leqq \frac{|z|}{1-|z|^2}$$

and

$$\frac{1-|z|^2}{(1+|z|^2)^3} \leqq |f'(z)| \leqq \frac{1+|z|^2}{(1-|z|^2)^3}$$

where the equalities are attained only by the so-called *Koebe functions*

$$f(z) = \frac{z}{(1-\varepsilon z)^2} \qquad (|\varepsilon| = 1).$$

Here, the functionals $\Omega[f]$ under consideration are the maximums and the minimums of $|f(z)|$ and $|f'(z)|$. Another famous distortion problem is the Bieberbach conjecture which asks, for the expansion of $f \in \mathcal{S}$

$$f(z) = z + \sum_{n=2}^{\infty} a_n z^n \qquad (|z| < 1),$$

whether the inequalities $|a_n| \leq n$ ($n = 2, 3, \ldots$) hold with the equalities again attained only by the Koebe function. This is a typical coefficient problem in which the functionals $\Omega[f]$ are the absolute values of the coefficients of the expansion. After many partial results, the Bieberbach conjecture was solved affirmatively by de Branges.

There are many other classical functionals $\Omega[f]$ and classes of functions on which $\Omega[.]$ are defined. In the present monograph, we are concerned with distortion problems for functionals of the form

$$\Omega[\mathcal{L} f] \qquad (f \in \mathcal{F}), \tag{1.1}$$

with $\Omega[.]$ being classical functionals, where $\mathcal{L}$ is a linear integral operator on $\mathcal{F}$ defined as follows. For the unit interval $I = [0, 1]$ on the real line, we take a probability measure σ supported on I, and identify it with the distribution function $\sigma(t)$. Thus $\sigma(t)$ is an increasing function on I such that $\sigma(0) = 0$ and $\sigma(1) = 1$. With this σ, we set

$$\mathcal{L} f(z) = \int_I \frac{f(zt)}{t} \, d\sigma(t) \qquad (f \in \mathcal{F}), \tag{1.2}$$

cf. Komatu [5, 6]; Komatu-Niino-Yang [1]. Thus $\mathcal{L}$ is parametrized by σ. In considering a distortion problem for a functional of the form (1.1), we vary σ in a class of probability measures, with $f \in \mathcal{F}$ fixed or f varying in $\mathcal{F}$.

The present monograph consists of two parts; I and II. In Part I, we study the dependence of our integral operator $\mathcal{L}$ on the probability measure σ. Starting from $\mathcal{L}$ in (1.2) with a measure σ prescribed, we first define the iterates $\mathcal{L}^n$ ($n = 0, 1, 2, \ldots$). This is done by the power series expansions of f and $\mathcal{L} f$. Interpolating $\mathcal{L}^n$ thus obtained, we next get positive powers $\mathcal{L}^\lambda$ ($\lambda > 0$). Under a moment condition on σ defining $\mathcal{L}$, the positive power $\mathcal{L}^\lambda$ has a representation of the form (1.2) with a probability measure σ_λ. Negative powers and complex powers of $\mathcal{L}$ are also discussed, but in a less general situation, as follows. Let us first restrict ourselves to the case where the measure σ has a density ρ, that is, $d\sigma(t) = \rho(t)\,dt$. We write the operator $\mathcal{L}$ as $\mathcal{L}[\rho]$ in order to emphasize the dependence on ρ or σ. Then, the composition $\mathcal{L}[\rho_1]\mathcal{L}[\rho_2]$ is realized by a multiplicative convolution of ρ_1 and ρ_2. Let us further specialize ourselves to the case $\sigma(t) = t^a$ with $a > 0$ fixed. Then we are reduced to a version of the standard fractional calculus, where usual differentiation is replaced by logarithmic one. Thus negative powers and complex powers of $\mathcal{L}$ for $\sigma(t) = t^a$ are defined. We hope that our general formulation (1.2) has examples admitting new fractional calculus.

In Part II of this monograph, we study distortion problems for functionals of the form (1.1) and related problems. Among these, a simple but typical result is given as follows. Consider a family of classical functionals

$$M_r[F] = \max_{|z|=r} |F(z)| \qquad (F \in \mathcal{F},\ 0 < r < 1).$$

Then the maximum principle yields $M_r[\mathcal{L} f] \leqq M_r[f]$, where the equality holds trivially when the probability measure σ determining $\mathcal{L}$ is the point mass at $t = 1$ defined by $\sigma^*(t) = 0$ for $0 \leqq t < 1$ and $\sigma^*(1) = 1$. Assuming $\sigma \neq \sigma^*$, we see by inspecting the proof that the equality $M_r[\mathcal{L} f] = M_r[f]$ for some r if and only if $f(z) = z$. Here, σ is arbitrary as far as $\sigma \neq \sigma^*$. Also, the extremal bound $M_r[f]$ for $M_r[\mathcal{L} f]$ is independent of the operator $\mathcal{L}$.

The function $f(z) = z$ also appears as the extremal function of distortion problems in which the classical functionals are given by

$$h(r) = \min_{|z|=r} \operatorname{Re} \frac{f(z)}{z}, \qquad H(r) = \max_{|z|=r} \operatorname{Re} \frac{f(z)}{z}$$

and $\Delta(r) = H(r) - h(r)$. For the classical functional $\Omega_r[f] = \Delta(r)$, the extremal bound for our functional $\Omega_r[\mathcal{L} f$

depends on the probability measure σ determining $\mathcal{L}$ as follows:

$$\Omega_r[\mathcal{L} f] \leq \frac{4}{\pi}\, \Omega_r[f] \int_I \arctan t \, d\sigma(t).$$

As extremal functions of our distortion problems, we encounter several elementary functions including the Koebe function, the identity function $f(z) = z$ and $\chi(z) = z/(1 - z)$. The function $\chi(z)$ is characterized by

$$f * \chi = \chi * f = f \qquad (f \in \mathcal{J}),$$

where $*$ stands for the Hadamard product defined by

$$\varphi * \psi(z) = \sum_{\nu=1}^{\infty} a_\nu b_\nu z^\nu$$

for $\varphi, \psi \in \mathcal{J}$ having the expansions

$$\varphi(z) = \sum_{\nu=1}^{\infty} a_\nu z^\nu, \qquad \psi(z) = \sum_{\nu=1}^{\infty} b_\nu z^\nu.$$

We also note that $\mathcal{L}$ in (1.2) always satisfies

$$\mathcal{L}(\varphi * \psi) = \varphi * \mathcal{L}\psi = \mathcal{L}\varphi * \psi.$$

For reader's convenience, we list here fundamental classes of functions which are frequently used in the text. Recall first that $\mathcal{J}$ is the totality of holomorphic functions $f(z)$ in the unit disk E such that $f(0) = 0$ and $f'(0) = 1$. A subclass of $\mathcal{J}$ is defined by

$$\mathcal{J}(\alpha) = \left\{ f \in \mathcal{J} \,\middle|\, \operatorname{Re} \frac{f(z)}{z} > \alpha \right\} \qquad \text{for } \alpha < 1.$$

We then set $\mathcal{J}^+ = \mathcal{J}(0)$ and $\mathcal{H} = \mathcal{J}(1/2)$. Sometimes it is convenient to consider $P(\alpha) = \{ z f(z) \mid f \in \mathcal{J}(\alpha)\}$, which has the name the *Carathéodory class* of order α. Recall next that $\mathcal{S}$ is the subclass of $\mathcal{J}$ consisting of univalent functions. We denote by $\mathcal{K}$ and $\mathcal{S}t$ the totalities of functions in $\mathcal{S}$ which are convex and starlike with respect to the origin, respectively.

PART I. INTEGRAL OPERATORS

In the present part, we shall deal with the linear integral operators playing the main roles of basic tools in constructing several functionals as the object of distortions which will be discussed in the latter part.

Beginning with the definition of the basic fundamental operator, we attach them properties of various kinds, and by specializing or generalizing them, we observe their several properties.

Chapter 1. Basic Integral Operators

§ 2. Linear integral operator

Now, we shall try to iterate the integral operator originally given by

$$(2.1) \qquad \mathcal{L}: \quad \mathcal{L}f(z) = \int_I \frac{f(zt)}{t}\, d\sigma(t)$$

in the unit disk E, where σ is a probability measure on the unit interval I, into the sequence

$$\{\mathcal{L}^n\}_{n=0}^{\infty},$$

such that the iterated sequence satisfies the aditivity condition

$$(2.2) \qquad \mathcal{L}^m \mathcal{L}^n = \mathcal{L}^{m+n}$$

for any non-negative integers m and n.

This iteration arises automatically. In fact, it is obtained by means of the recurrence formula in the induction:

$$\mathcal{L}^0 = \mathrm{id}, \quad \mathcal{L}^n = \mathcal{L}\mathcal{L}^{n-1} \quad (n \geqq 1).$$

In parallel with the integral operators, the method due to power series

$$(2.3)\qquad f(z) = \sum_{\nu=1}^{\infty} c_\nu z^\nu, \qquad c_1 = 1,$$

is also often shown to be a useful auxiliary one, as will be seen everywhere.

Let $\mathcal{F}$ denote the class of analytic functions f which are holomorphic in E and normalized by

$$(2.4)\qquad f(0) = f'(0) - 1 = 0.$$

Then a linear integral operator $\mathcal{L}$ is defined by (2.1).

Let the Taylor expansion of $f \in \mathcal{F}$ be given by (2.3). Then, substitution of (2.3) into (2.1) followed by termwise integration yields

$$\mathcal{L} f(z) = \sum_{\nu=1}^{\infty} \alpha_\nu c_\nu z^\nu,$$

where $\{\alpha_\nu\}_{\nu=1}^{\infty}$ is the moment sequence with respect to σ defined by

$$(2.5)\qquad \alpha_\nu = \int_I t^{\nu-1} d\sigma(t) \qquad (\nu = 1, 2, \ldots).$$

As immediately seen from its expression, the sequence (2.5) is always decreasing and non-negative; in particular,

$$\alpha_\nu = 1.$$

§ 3. Additive family of operators

In the present section, we shall attempt to interpolate the class $\{\mathcal{L}^n\}_{n=0}^{\infty}$ into a family $\{\mathcal{L}^{\lambda}\}_{\lambda\geqq 0}$ depending on a continuous parameter λ.

In general, concerning the integral operator of the form (2.1), it is readily seen that $f \in \mathcal{F}$ implies $\mathcal{L} f \in \mathcal{F}$. In fact, since $f(0) = 0$, $f'(0) = 1$, we have

$$\mathcal{L} f(0) = \int_I \frac{f(0)}{t}\, d\sigma(t) = 0,$$

$$\mathcal{L} f'(0) = \int_I f'(0)\, d\sigma(t) = 1.$$

As already mentioned, the iteration $\{\mathcal{L}^n\}_{n=0}^{\infty}$ arises automatically by

$$\mathcal{L}^0 = \mathrm{id}, \quad \mathcal{L}^n = \mathcal{L}\mathcal{L}^{n-1} \quad (n = 1, 2, \dots)$$

or

$$\mathcal{L}^n f(z) = \sum_{\nu=1}^{\infty} \alpha_\nu^{\,n} c_\nu z^\nu. \tag{3.1}$$

Now, the problem of interpolation is to determine a family

$\{\mathcal{L}^{\lambda}\}_{\lambda}$ depending on a non-negative continuous parameter λ in such a way that the additivity

$$\mathcal{L}^{\lambda}\mathcal{L}^{\mu} = \mathcal{L}^{\lambda+\mu} \tag{3.2}$$

remains valid; namely, the family possesses the structure of semiring.

Under the condition that $f \in \mathcal{F}$ implies always $\mathcal{L}^{\lambda} f \in \mathcal{F}$ we suppose here

$$\mathcal{L}^{\lambda} f(z) = \sum_{\nu=1}^{\infty} \gamma_{\nu}(\lambda) c_{\nu} z^{\nu},$$

where, in particular,

$$\gamma_{\nu}(n) = \alpha_{\nu}^{\,n} \qquad (\nu = 1, 2, \ldots; \; n = 0, 1, \ldots).$$

Then, by observing a particular function

$$f(z) = \frac{z}{1-z} \in \mathcal{F},$$

for instance, we get from (3.2) the functional equation

$$\gamma_{\nu}(\lambda)\, \gamma_{\nu}(\mu) = \gamma_{\nu}(\lambda + \mu).$$

Under a very weak supposition on γ_{ν}, for instance, its boundedness in any small interval, the solution of this equation is given by

$$\tau_{\nu}(\lambda) = \alpha_{\nu}^{\lambda}. \tag{3.3}$$

We now proceed to state theorems concerning the interpolation.

THEOREM 3.1. *If* $\{\alpha_{\nu}\}$ *is any sequence which satisfies* $\alpha_1 = 1$ *and* $\limsup |\alpha_{\nu}|^{1/\nu} \leq 1$, *then the family of operators* $\{\mathcal{L}^{\lambda}\}$ *depending on a continuous parameter* λ *which is defined by*

$$\mathcal{L}^{\lambda} f(z) := \sum_{\nu=1}^{\infty} \alpha_{\nu}^{\lambda} c_{\nu} z^{\nu} \tag{3.4}$$

where $f(z) = \Sigma_{\nu=1}^{\infty} c_{\nu} z^{\nu} \in \mathcal{J}$ *satisfies the additivity* (3.2)

Proof. The assertion is readily seen. In particular, the assumption $\limsup |\alpha_{\nu}|^{1/\nu} \leq 1$ together with $\alpha_1 = 1$ ensures that (3.4) belongs to $\mathcal{J}$. □

Now, we shall be concerned ourselves about the integral representation of the the form for the operator $\mathcal{L}^{\lambda}$. It is just connected with a moment problem of Hausdorff type and hence it refers to the notion of total monotonicity (totale Monotonie).

In general, let a sequence $\{\mu_\nu\}_{\nu=0}^{\infty}$ be given. Then, corresponding to every polynomial

$$p(t) = \sum_{\nu=0}^{n} p_\nu t^\nu,$$

we observe a number

$$Mp = Mp(t) = \sum_{\nu=0}^{n} p_\nu \mu_\nu,$$

which is called the moment of p.

The moment problem is to obtain a function σ of bounded variation satisfying

$$\int_I p(t)\, d\sigma(t) = Mp$$

or, equivalently

$$\int_I t^\nu d\sigma(t) = \mu_\nu \quad (\nu = 0, 1, \ldots). \tag{3.5}$$

In general, a real sequence $\{\mu_\nu\}_{\nu=1}^{\infty}$ is called totally monotone if it satisfies

$$\Delta^\kappa \mu_\nu = \sum_{j=0}^{\kappa} (-1)^j \binom{\kappa}{j} \mu_{\nu+j} \geq 0 \tag{3.6}$$

$$(\kappa = 0, 1, \ldots; \nu = 1, 2, \ldots),$$

where

$$\Delta^0 \mu_\nu = \mu_\nu \text{ and } \Delta^\kappa \mu_\nu = \Delta^{\kappa-1} \mu_\nu - \Delta^{\kappa-1} \mu_\nu \quad (\kappa \geqq 1).$$

We state here the Hausdorff's theorem showing the condition on solvability of the moment problem; Hausdorff [1,2]; cf. also Shohat and Tamarkin [1].

THEOREM 3.2. *In order that the moment problem* (3.5) *has a monotone function* $\sigma(t)$ *as a solution, it is necessary and sufficient that the sequence of moments is totally monotone.*

Proof. We first notice that for the existence of a solution it is necessary that any non-negative function must have a non-negative moment. In particular, $t^\nu (1 - t)^\kappa$ $(\kappa, \nu = 0, 1, \ldots)$ has a non-negative moment; namely, it satisfies (3.6). In order to verify the sufficiency part of Theorem, we first show that, in view of (3.6), every non-negative polynomial has a non-negative moment. For that purpose, we write a non-negative polynomial

$$p(t) = \sum_{j=0}^{k} p_j t^j$$

in the form

$$p(t) = \sum_{\nu=0}^{\kappa} a_\nu \binom{\kappa}{\nu} t^\nu (1 - t)^{\kappa-\nu} \tag{3.7}$$

uniquely so far as $\kappa \geqq k$. Its coefficients dependent on n are given by

$$(3.8) \qquad a_\nu = \sum_{j=0}^{k} p_j \frac{\nu!(\kappa - j)!}{\kappa!(\nu - j)!}$$

$$= \sum_{j=0}^{k} p_j \frac{\nu(\nu - 1)\ldots(\nu - j + 1)}{\kappa(\kappa - 1)\ldots(\kappa - j + 1)} .$$

To verify this, we first get from

$$\sum_{j=0}^{k} p_j t^j = \sum_{\nu=0}^{\kappa} a_\nu \binom{\kappa}{\nu} t^\nu (1 - t)^{\kappa-\nu}$$

$$= \sum_{\nu=0}^{\kappa} a_\nu \binom{\kappa}{\nu} t^\nu \sum_{\iota=0}^{\kappa-\nu} \binom{\kappa - \nu}{\iota} (- t)^\iota$$

$$= \sum_{\nu=0}^{\kappa} a_\nu \binom{\kappa}{\nu} \sum_{j=\nu}^{\kappa} (- 1)^{j-\nu} \binom{\kappa - \nu}{j - \nu} t^j$$

$$= \sum_{j=0}^{\kappa} \sum_{\nu=0}^{j} (- 1)^{j-\nu} \binom{\kappa}{\nu} \binom{\kappa - \nu}{j - \nu} a_\nu t^j ,$$

the relation

$$(3.9) \qquad p_j = \sum_{\nu=0}^{j} (- 1)^{j-\nu} \binom{\kappa}{\nu} \binom{\kappa - \nu}{j - \nu} a_\nu .$$

In order to obtain the inversion formula, we notice

$$(1-1)^{\nu-\iota} = \sum_{i=0}^{\nu-\iota} \frac{(\nu-\iota)!}{(\nu-i-\iota)!\,i!}(-1)^{i}$$

from which we derive

$$a_{\nu} = \sum_{\iota=0}^{n} \frac{\nu!}{\iota!(\nu-\iota)!}\, a_{\iota} \sum_{i=0}^{n-\iota} \frac{(\nu-\iota)!}{(\nu-i-\iota)!\,i!}(-1)^{i}$$

$$= \sum_{\iota=0}^{n} \sum_{j=\iota}^{n} (-1)^{j-\iota} \frac{\nu!(\kappa-j)!}{\kappa!(\nu-j)!} \binom{\kappa}{\iota}\binom{\kappa-\iota}{j-\iota} a_{\iota}$$

$$= \sum_{j=0}^{n} \frac{\nu!(\kappa-j)!}{\kappa!(\nu-j)!} \sum_{\iota=0}^{j} (-1)^{j-\iota} \binom{\kappa}{\iota}\binom{\iota-\iota}{j-\iota} a_{\iota}.$$

Compared with (3.9), we have shown that (3.8) really holds. Introducing the polynomial

$$p_{\kappa}(t) := \sum_{j=0}^{n} p_{j} \frac{\kappa t(\kappa t-1)\dots(\kappa t-j+1)}{\kappa(\kappa-1)\dots(\kappa-j+1)},$$

the relation can be also expressed by

$$a_{\nu} = p_{\kappa}\left(\frac{\nu}{\kappa}\right).$$

Since $(\kappa t-j)/(\kappa-j) \to t$ and hence $p_{\kappa}(t) \to p(t)$ as $\kappa \to \infty$ uniformly in I, we see that for sufficiently large κ we have

$$\left|a_{\nu} - p\left(\frac{\nu}{\kappa}\right)\right| < \varepsilon \qquad (\nu = 0, 1, \dots, \kappa).$$

Therefore, if the polynomial p remains positive in I, then (3.7) has positive coefficients alone. Now, if we put

$$\lambda_{\nu,\kappa} := \binom{\kappa}{\nu} \Delta^{\kappa-\nu} \mu_\nu$$

for the sake of brevity, we see from (3.6) that

$$M\,p = \sum_{\nu=0}^{\kappa} a_\nu \lambda_{\nu,\kappa} .$$

Hence, if p is totally monotone, every polynomial with non-negative coefficients has a non-negative moment. Further, if ν is large enough such that $|a_\nu - p(\nu/\kappa)| < \varepsilon$, then

$$\left| M\,p - \sum_{\nu=0}^{\kappa} p\left(\frac{\nu}{\kappa}\right) \lambda_{\nu,\kappa} \right| \leqq \varepsilon \sum_{\nu=0}^{\kappa} \lambda_{\nu,\kappa} = \varepsilon \mu_0 .$$

Consequently, for every polynomial we have

$$\lim_{\kappa\to\infty} \sum_{\nu=0}^{\kappa} p\left(\frac{\nu}{\kappa}\right) \lambda_{\nu,\kappa} = M\,p . \tag{3.10}$$

We shall verify that the last relation (3.10) yields a solution of the moment problem by approximating step functions. For that purpose, let $\sigma_\kappa(t)$ be constant except jumps $\lambda_{\kappa,\nu}$ at ν/κ ($\nu = 0, 1, \ldots, \kappa$), respectively, and satisfy the requirement of arithmetic mean at jump points; more precisely,

$$\sigma_\kappa(0) = 0, \qquad \sigma_\kappa(1) = \sum_{\nu=0}^{\kappa} \lambda_{\kappa,\nu} = \mu_0,$$

$$\sigma_\kappa(t-0) = \sum_{\nu/\kappa < t} \lambda_{\kappa,\nu}, \quad \sigma_\kappa(t+0) = \sum_{\nu/\kappa \leq t} \lambda_{\kappa,\nu},$$

$$\sigma_\kappa(t) = \frac{\sigma_\kappa(t-0) + \sigma_\kappa(t+0)}{2}$$

in $0 < t < 1$. In terms of this monotone function, the relation (3.8) is expressed in the form

$$\lim_{\kappa\to\infty} \int_I p(t)\, d\sigma_\kappa(t) = Mp. \tag{3.11}$$

By means of $0 \leq \sigma_\kappa(t) \leq \mu_0$, the monotone sequence $\{\sigma_\kappa\}$ is uniformly bounded and hence, in view of Helly's selection theorem, we can choose a subsequence $\{\sigma_{n_\kappa}\}$ which converges everywhere. Its limit function

$$\hat{\sigma}(t) = \lim \sigma_{n_\kappa}(t)$$

is again monotone and attains the values 0, μ_0 at the points 0, 1, respectively. However, it may not satisfy the assertion of arithmetic mean at jump points. In view of the uniform boundedness of $\{\sigma_\kappa\}$, for every polynomial

$$\int p'(t)\sigma_\kappa(t)\,dt \quad \text{converges to} \quad \int p'(t)\hat{\sigma}(t)\,dt$$

and since we have by means of integration by parts

$$\int_I p(t)\,d\hat{\sigma}(t) = p(1)\hat{\sigma}(1) - \int_I p'(t)\hat{\sigma}(t)\,dt,$$

we see that

$$\int_I p(t)\,d\sigma_\kappa(t) \quad \text{converges to} \quad \int_I p(t)\,d\hat{\sigma}(t).$$

Consequently, in view of (3.11), $\hat{\sigma}$ is a solution of the moment problem. On the other hand, we know that under the the assertion of arithmetic mean there exists a unique solution. Hence, at every continuity point of σ we have

$$\hat{\sigma}(t) = \lim \sigma_{n_\kappa}(t) = \sigma(t),$$

namely, the sequence $\{\sigma_{n_\kappa}\}$ contains a subsequence which converges to σ. Thus, the proof has been completed. □

By means of Theorem 3.2 just verified, if $\{\alpha_\nu^\lambda\}_\nu$ is totally monotone, then the moment problem

$$\int_I t^{\nu-1}\,d\sigma(t) = \alpha_\nu^\lambda \qquad (\nu = 1, 2, \ldots)$$

has a solution σ_λ. The solution is then unique and in view of $\alpha_1^\lambda = 1$ it is a probability measure. Finally, we have

$$\mathcal{L}^{\lambda} f(z) = \sum_{\nu=1}^{\infty} \alpha_{\nu}^{\lambda} c_{\nu} z^{\nu}$$

$$= \int_{I} \sum_{\nu=1}^{\infty} c_{\nu} z^{\nu} t^{\nu-1} d\sigma_{\lambda}(t)$$

$$= \int_{I} \frac{f(zt)}{t} d\sigma_{\lambda}(t).$$

REMARK. It is known that the solution of moment problem may be explicitly represented in the form

$$\sigma_{\lambda}(t) = \sum_{j=0} (2j+1) q_{j} \int_{0}^{t} P_{j}(2\tau - 1) d\tau,$$

where P_j is the Legendre polynomial of order j and q_j denotes the value

$$q_j = \int_0^1 P_j(2\tau - 1) d\sigma_{\lambda}(\tau)$$

in which

$$\int_0^1 \tau^{\nu-1} d\sigma_{\lambda}(\tau)$$

is to be replaced by α_{ν}^{λ}; in particular, q_j is a definite linear combination of $\{\alpha_{\nu}^{\lambda}\}_{\nu=1}^{j+1}$.

There are several sequences $\{\alpha_\nu\}_{\nu=1}^{\infty}$ with $\alpha_1 = 1$ which are totally monotone. Among them there are two extreme cases. One is the case where $\alpha_1 = 1$ and $\alpha_\nu = 0$ for $\nu > 1$ while another is the case where $\alpha_\nu = 1$ for every ν. They correspond to the operators $\mathcal{L}$ satisfying

$$\mathcal{L} f(z) = z \quad \text{and} \quad \mathcal{L} f(z) = f(z)$$

for every $f \in \mathcal{F}$, respectively. In terms of probability measure, they correspond to $\sigma(t)$ concentrated at $t = 0$ and $t = 1$, namely, they are the Dirac measure at $t = 0$ and $= 1$, respectively. They play exceptional roles also for the continuity for $\lambda \to +0$ and the limit behavior as $\lambda \to \infty$, as shown in the following theorem:

THEOREM 3.3. *The limit relations*

$$\lim_{\lambda\to+0} \mathcal{L}^{\lambda} f(z) = f(z) \quad \textit{and} \quad \lim_{\lambda\to\infty} \mathcal{L}^{\lambda} f(z) = z$$

hold for every $f \in \mathcal{F}$ *in* E *uniformly in the wider sense except the extreme cases where* $\mathcal{L} f(z) = z$ *for the former relation and* $\mathcal{L} f(z) = f(z)$ *for the latter.*

Proof. The assertion follows from the series expansion

$$\mathcal{L}^{\lambda} f(z) = \sum_{\nu=1}^{\infty} \alpha_\nu^{\lambda} c_\nu z^{\nu}.$$

In fact, except the case where $\mathcal{L} f(z) = z$, that is, $\alpha_1 = 1$ and $\alpha_\nu = 0$ for $\nu > 1$, the sequence $\{\alpha_\nu\}$ satisfies $0 < \alpha_\nu \leq \alpha_1 = 1$ for every $\nu > 1$ whence follows

$$\mathcal{L}^{\lambda} f(z) \to f(z) \quad (\lambda \to +0).$$

On the other hand, except the case where $\mathcal{L} f(z) = f(z)$, that is, $\alpha_\nu = 1$ for every ν, namely, $\alpha_\nu = 1$ for every ν, the sequence $\{\alpha_\nu\}$ satisfies $0 \leq \alpha_\nu < \alpha_1 = 1$ for every $\nu > 1$, whence follows

$$\mathcal{L}^{\lambda} f(z) \to z \quad (\lambda \to \infty).$$

Every limit is uniform in the wider sense. □

The first relation in Theorem 3.4 states that the operator family $\{\mathcal{L}^{\lambda}\}_{\lambda}$ under consideration shows the continuity as $\lambda \to 0$, while the second limit relation shows that it possesses the tendency of "rounding" the image as λ increases towards ∞, in other words, the image of E by $w = \mathcal{L}^{\lambda} f(z)$ becomes gradually round as λ tends to ∞, unless $\mathcal{L} f(z) = f(z)$ for every $f \in \mathcal{F}$.

We now observe the Hadamard product. The Hadamard product of two power series

$$\varphi(z) = \sum_{\nu=1}^{\infty} a_\nu z^\nu, \qquad \psi(z) = \sum_{\nu=1}^{\infty} b_\nu z^\nu \tag{3.12}$$

is defined by

$$\varphi * \psi(z) = \sum_{\nu=1}^{\infty} a_\nu b_\nu z^\nu. \tag{3.13}$$

It is readily seen that $\varphi, \psi \in \mathcal{J}$ implies $\varphi * \psi \in \mathcal{J}$, $\varphi * \psi = \psi * \varphi$ and the particular function

$$\chi(z) = \frac{z}{1-z} = \sum_{\nu=1}^{\infty} z^{\nu} \tag{3.14}$$

plays a role of unit function with respect to the operation $*$ in the class $\mathcal{J}$; namely,

$$f * \chi = \chi * f = f \quad (f \in \mathcal{J}).$$

THEOREM 3.4. *For any pair of functions* (3.8), *any operator* $\mathcal{L}$ *under consideration satisfies*

$$\mathcal{L}(\varphi * \psi) = \varphi * \mathcal{L}\psi = \mathcal{L}\varphi * \psi.$$

Proof. Direct calculation yields

$$\begin{aligned}
\mathcal{L}(\varphi * \psi)(z) &= \int_I \frac{(\varphi * \psi)(zt)}{t}\, d\sigma(t) \\
&= \int_I \left(\varphi(z) * \frac{\psi(zt)}{t} \right) d\sigma(t) \\
&= \varphi(z) * \int_I \frac{\psi(zt)}{t}\, d\sigma(t) \\
&= (\varphi * \mathcal{L}\psi)(z).
\end{aligned}$$

The remaining part is seen from

$$\mathcal{L}\varphi * \psi = \psi * \mathcal{L}\varphi = \mathcal{L}(\psi * \varphi) = \mathcal{L}(\varphi * \psi). \qquad \square$$

COROLLARY 3.1. *The action of* $\mathcal{L}$ *on* $f \in \mathcal{F}$ *is reduced to the Hadamard product of* f *with* $\mathcal{L}\chi$ *where* χ *denotes the definite function* (3.14).

Proof. By means of the relation in Theorem 3.4, we get

$$\mathcal{L} f = \mathcal{L}(f * \chi) = f * \mathcal{L}\chi. \qquad \square$$

§ 4. The case possessing a density

We now suppose that a probability measure possesses a density. In such a case, the density ρ is given by mean of the measure σ in the form

$$(4.1) \qquad \rho(t) = \int_0^t \rho(\tau)\, d\tau, \quad \rho(\tau) \geqq 0, \quad \int_I \rho(\tau)\, d\tau = 1.$$

The operator generated by this measure will be denoted by $\mathcal{L}[\rho]$:

$$(4.2) \qquad \mathcal{L}[\rho] f(z) := \int_I \frac{f(zt)}{t} \rho(t)\, dt$$

for the following discussions; cf. Komatu [14].

For the purpose to later sections, we begin with the following Lemma:

LEMMA 4.1. *The product of two operators* $\mathcal{L}[p]$ *and* $\mathcal{L}[q]$ *becomes*

$$\mathcal{L}[p]\,\mathcal{L}[q] = \mathcal{L}[\rho]$$

where

$$\rho(t) = \int_t^1 p(s)\,q\left(\frac{t}{s}\right)\frac{ds}{s}. \tag{4.3}$$

Proof. Direct calculation shows that

$$\begin{aligned}
\mathcal{L}[p]\mathcal{L}[q]f(z) &= \int_I p(s)\frac{ds}{s}\int_I f(zs\tau)\,q(\tau)\frac{d\tau}{\tau} \\
&= \int_I p(s)\frac{ds}{s}\int_0^s f(zt)\,q\left(\frac{t}{s}\right)\frac{dt}{t} \\
&= \int_I f(zt)\frac{dt}{t}\int_t^1 p(s)\,q\left(\frac{t}{s}\right)ds \\
&= \int_I \frac{f(zt)}{t}\,\rho(t)\,dt = \mathcal{L}[\rho]f(z)
\end{aligned}$$

with ρ stated in (4.3). □

REMARK. If we put $t = e^{-v}$ and accordingly $p(t) = P(v)$, $q(t) = Q(v)$ and $\rho(t) = R(v)$, then the expression in the Lemma 4.1 becomes

$$R(v) = \int_0^v P(u)\,Q(v-u)\,du.$$

This shows that R is the convolution of P and Q:

$$R = P * Q.$$

LEMMA 4.2. *Let every member $\mathcal{L}^{\lambda}$ of the family $\{\mathcal{L}^{\lambda}\}_{\lambda>0}$ be generated by a measure which possesses the density* ρ_{λ}: $\mathcal{L}^{\lambda} = \mathcal{L}[\rho_{\lambda}]$. *Then the additivity* $\mathcal{L}^{\lambda}\mathcal{L}^{\mu} = \mathcal{L}^{\lambda+\mu}$ *is chacterized by*

$$\int_t^1 \rho_{\lambda}(s)\rho_{\mu}\left(\frac{t}{s}\right)\frac{ds}{s} = \rho_{\lambda+\mu}(t). \tag{4.4}$$

Proof. In view of Lemma 4.1, we have $\mathcal{L}[\rho_{\lambda}]\mathcal{L}[\rho_{\mu}] = \mathcal{L}[\rho]$ with $\rho(t)$ given by the expression of the left-hand member in (4.4). Hence the additivity is characterized by the condition that $\mathcal{L}[\rho]f(z) = \mathcal{L}[\rho_{\lambda+\mu}]f(z)$ holds for any $f \in \mathcal{F}$. This condition applied, for instance, to a particular function

$$f(z) = \frac{z}{1-z} \in \mathcal{F}$$

yields, by comparing the coefficients of z^{ν},

$$\int_I t^{\nu-1}\rho(t)\,dt = \int_I t^{\nu-1}\rho_{\lambda+\mu}(t)\,dt.$$

In view of the uniqueness of the solution of ordinary moment problem, we obtain

$$\rho = \rho_{\lambda+\mu}.$$

Conversely, if $\rho = \rho_{\lambda+\mu}$, it is evident that the additivity holds. □

§ 5. Operator generated by t^a

In the present section, we shall observe the probability measure σ depending on a parameter a which is defined by

$$(5.1) \qquad \sigma(t\,;\,a) = t^a \quad (a > 0);$$

cf. Komatu [14].

We first show in the following Theorem that with respect to (5.1) the measure $\sigma_\lambda(t\,;\,a)$ as well as the operator generated by this measure are obtained in the explicit forms. Accordingly, this measure will be in the subsequent lines referred to as an illustrating example very often.

THEOREM 5.1. *The additive family of operators generated by (5.1) is given by the measure* $\sigma_\lambda(t\,;\,a)$ *with the density* $\rho_\lambda(t\,;\,a)$ *defined by*

$$\sigma_\lambda(t\,;\,a) = \int_0^t \rho_\lambda(\tau;\,a)\,d\tau,$$

$$\rho_\lambda(t\,;\,a) = \frac{a^\lambda}{\Gamma(\lambda)}\, t^{a-1}\left(\log\frac{1}{t}\right)^{\lambda-1}.$$

Proof. The condition stated in Lemma 4.2 can be verified by direct calculation. In fact, we have

$$\int_t^1 \rho_\lambda(s\,;\,a)\rho_\mu\left(\frac{t}{s}\,;\,a\right)\frac{ds}{s}$$

$$= \frac{a^{\lambda+\mu}}{\Gamma(\lambda)\Gamma(\mu)} \int_t^1 s^{a-1}\left(\log\frac{1}{s}\right)^{\lambda-1}\left(\frac{t}{s}\right)^{a-1}\left(\log\frac{s}{t}\right)^{\mu-1}\frac{ds}{s}$$

$$= \frac{a^{\lambda+\mu}}{\Gamma(\lambda)\Gamma(\mu)}\, t^{a-1}\int_t^1 \left(\log\frac{1}{s}\right)^{\lambda-1}\left(\log\frac{s}{t}\right)^{\mu-1}\frac{ds}{s}$$

$$\left[\log\frac{1}{s} = u\log\frac{1}{t}\right]$$

$$= \frac{a^{\lambda+\mu}}{\Gamma(\lambda)\Gamma(\mu)}\, t^{a-1}\left(\log\frac{1}{t}\right)\int_I u^{\lambda-1}(1-u)^{\mu-1}du$$

$$= \frac{a^{\lambda+\mu}}{\Gamma(\lambda)\Gamma(\mu)}\, t^{a-1}\left(\log\frac{1}{t}\right)^{\lambda+\mu-1}\frac{\Gamma(\lambda)\Gamma(\mu)}{\Gamma(\lambda+\mu)} = \rho_{\lambda+\mu}(t\,;\,a).$$

The assertion may be alternatively verified as follows:

In fact, since the moment with respect to the measure $\sigma(t\,;\,a)$ is equal to

$$\alpha_\nu(a) = \int_I t^{\nu-1}\,d\sigma(t\,;\,a) = \frac{a}{\nu+a-1}.$$

It is sufficient to show that the moment with respect to the measure $\sigma_\lambda(t\,;\,a)$ mentioned in the Theorem 5.1 is equal to $\alpha_\nu(a)^\lambda$, what is an immediate consequence of a familiar formula

$$\int_I t^{\kappa-1}\left(\log\frac{1}{t}\right)^{\lambda-1} dt = \frac{\Gamma(\lambda)}{\kappa^\lambda}.$$
□

It is noted, in passing, that the total monotonicity of $\{(a/(\nu + a - 1))^\lambda\}_\nu$ is dirctly checked beforehand by taking notice of the last relation. In fact, we have

$$\Delta^\kappa\left(\frac{a}{\nu + a - 1}\right)^\lambda$$

$$= \frac{1}{\Gamma(\lambda)}\int_I t^{(\nu-1)/a}(1 - t)^\kappa\left(\log\frac{1}{t}\right)^{\lambda-1} dt \geqq 0.$$

According to Theorem 5.1, we shall denote in the following lines $\mathcal{L}[\sigma_\lambda(t\ ;\ a)]$ briefly by $\mathcal{L}(a)^\lambda$:

$$(5.2)\qquad \mathcal{L}(a)^\lambda f(z) := \frac{a^\lambda}{\Gamma(\lambda)}\int_I f(zt)\, t^{a-2}\left(\log\frac{1}{t}\right)^{\lambda-1} dt.$$

The behavior of the general family $\{\mathcal{L}^\lambda\}$ as $\lambda \to +0$ and $\lambda \to \infty$ has been shown in Theorem 3.4. However, in case of $\sigma(t\ ;\ a)$, since the extreme cases do not appear, we can mention the following theorem:

THEOREM 5.2. *The limit relations*

$$\lim_{\lambda\to+0} \mathcal{L}(a)^\lambda f(z) = f(z) \quad \textit{and} \quad \lim_{\lambda\to\infty} \mathcal{L}(a)^\lambda f(z) = z$$

hold for every $f \in \mathcal{J}$ *uniformly in* E *in the wider sense.*

On the other hand, the behavior as $a \to +0$ and $a \to \infty$ becomes as follows:

THEOREM 5.3. *The limit relations*

$$\lim_{a \to +0} \mathcal{L}(a)^{\lambda} f(z) = z \quad \textit{and} \quad \lim_{a \to \infty} \mathcal{L}(a)^{\lambda} f(z) = f(z)$$

hold for every $f \in \mathcal{F}$ *in* E *uniformly in the wider sense.*

Proof. Let z be restricted on any compact set in E. Then both $|f(zt)/t - z|$ and $|f(zt)/t - f(z)|$ possess for every $t \in I$ a bound M, say. First, we have

$$\mathcal{L}(a)^{\lambda} f(z) - z$$

$$= \frac{a^{\lambda}}{\Gamma(\lambda)} \int_I \left(\frac{f(zt)}{t} - z \right) t^{a-1} \left(\log \frac{1}{t} \right)^{\lambda-1} dt.$$

For any $\varepsilon > 0$ there exists a $\tau \in (0, 1)$ such that $|f(zt)/t - z| < \varepsilon/2$ as $0 \leqq t < \tau$, and hence for $a < 1$

$$|\mathcal{L}(a)^{\lambda} f(z) - z| < \frac{\varepsilon}{2} \frac{a^{\lambda}}{\Gamma(\lambda)} \int_0^{\tau} t^{a-1} \left(\log \frac{1}{t} \right)^{\lambda-1} dt$$

$$+ M\tau^{a-1} \frac{a^{\lambda}}{\Gamma(\lambda)} \int_{\tau}^{1} \left(\log \frac{1}{t} \right)^{\lambda-1} dt.$$

The first summand of this estimate is always less than $\varepsilon/2$, while the second summand becomes less than $\varepsilon/2$ provided a is sufficiently near to zero. It leads to the first relation in the Theorem. Next, we have

$$\mathcal{L}(a)^{\lambda} f(z) - f(z)$$

$$= \frac{a^{\lambda}}{\Gamma(\lambda)} \int_I \left(\frac{f(zt)}{t} - f(z) \right) t^{a-1} \left(\log \frac{1}{t} \right)^{\lambda-1} dt .$$

For any $\varepsilon > 0$ there exists a $\tau \in (0, 1)$ such that $| f(zt)/t - f(z)| < \varepsilon/2$ as $1 - \tau < t \leqq 1$, and hence for $a > 1$

$$|\mathcal{L}^{\lambda} f(z) - f(z)|$$

$$< M(1 - \tau)^{a-1} \frac{a^{\lambda}}{\Gamma(\lambda)} \int_0^{1-\tau} \left(\log \frac{1}{t} \right)^{\lambda-1} dt$$

$$+ \frac{\varepsilon}{2} \frac{a^{\lambda}}{\Gamma(\lambda)} \int_{1-\tau}^{1} t^{a-1} \left(\log \frac{1}{t} \right)^{\lambda-1} dt .$$

Since the second summand of this estimate is always less than $\varepsilon/2$ while the first summand becomes less than $\varepsilon/2$ for a large enough, the second relation follows.

Though the proof given here has been based on the integral reprentation for $\mathcal{L}(a)^{\lambda} f(z)$, a rather brief proof may be given by referring to its series expansion. □

In this occasion, we supplement here a remark.

The discussions developed for the case generated by the measure

$$\rho(t ; a) = at^{a-1}, \quad \rho_{\lambda}(t ; a) = \frac{a^{\lambda}}{\Gamma(\lambda)} t^{a-1} \left(\log \frac{1}{t} \right)^{\lambda-1}$$

will be formally generalized into the case

$$\rho(t;\, a, b) = \frac{a^b}{\Gamma(b)}\, t^{a-1}\left(\log\frac{1}{t}\right)^{b-1}.$$

However, the latter can be reduced essentially to the former. In fact, we have only to take into account the relation

$$\rho_\lambda(t;\, a, b) = \rho_{b+\lambda}(t;\, a, 1) = \rho_{b+\lambda}(t;\, a).$$

On the other hand, it may be remarked that the operator $\mathcal{L}(1)$ was observed by Srivastava-Owa [1], $\mathcal{L}(2)$ by Libera [1] and Livingston [1], and that $\mathcal{L}(k)$ for an integer $k > 1$ was studied by Bernardi [1], all in connection with certain subclasses of univalent functions.

Chapter 2. Properties of Integral Operators

§ 6. Relations to fractional calculus

Our integral operators under considerations are closely connected with the fractional calculus, which will be investigated in the present section; cf. Komatu [14].

We first observe, in general, the integral operator $\mathcal{L}^{\lambda}$ given by

$$\mathcal{L}^{\lambda} f(z) = \int_I \frac{f(zt)}{t}\, d\sigma_{\lambda}(t),$$

and notice that it commutes the differentiation operator

$$\theta = \frac{d}{d \log z}; \tag{6.1}$$

cf. also Komatu [5].

THEOREM 6.1. *In genaral, we have for any* $\lambda \geqq 0$

$$\theta \mathcal{L}^{\lambda} = \mathcal{L}^{\lambda} \theta . \tag{6.2}$$

Proof. By differentiating the expression of $\mathcal{L}^{\lambda} f(z)$, we obtain for $\lambda \geqq 0$

$$\theta \mathcal{L}^{\lambda} f(z) = \frac{d}{d \log z} \int_I \frac{f(zt)}{t} \, d\sigma_{\lambda}(t)$$

$$= z \int_I f'(zt) \, d\sigma_{\lambda}(t)$$

$$= \int_I \frac{zt f'(zt)}{t} \, d\sigma_{\lambda}(t)$$

$$= \mathcal{L}^{\lambda}(zf'(z))$$

$$= \mathcal{L}^{\lambda} \theta f(z).$$

An alternative proof by series approach may be more simple. In fact, since

$$f(z) = \sum_{\nu=1}^{\infty} c_{\nu} z^{\nu}$$

is transformed into

$$\mathcal{L}^{\lambda} f(z) = \sum_{\nu=1}^{\infty} \alpha_{\nu}^{\lambda} c_{\nu} z^{\nu},$$

it is readily seen that the operations $d/d \log z$ and $\mathcal{L}^{\lambda}$ are commutative. □

For a while, we restrict ourselves to a distinguished family of operators $\{\mathcal{L}(1)^{\lambda}\}$ which is generated by a special probability measure

$$\sigma(t) = t.$$

Then, we can introduce the complex integral form for $\mathcal{L}(1)^{\lambda}$. By definition, we get in fact

$$\mathcal{L}(1)^{\lambda} f(z) = \frac{1}{\Gamma(\lambda)} \int_I \frac{f(zt)}{t} \left(\log \frac{1}{t}\right)^{\lambda-1} dt$$

$$\left[t = \frac{\zeta}{z} \right]$$

$$= \frac{1}{\Gamma(\lambda)} \int_0^z \frac{f(\zeta)}{\zeta} \left(\log \frac{z}{\zeta}\right)^{\lambda-1} d\zeta,$$

where the integration is taken along the ray from 0 to z or along any homotopic curve in the unit disk punctured at the origin and the kernel of the integral denotes the branch which is real and positive on the ray.

On the other hand, we have considered the differential operator θ defined by

$$\theta = \frac{d}{d \log z} = z \frac{d}{dz}.$$

The inverse operator θ^{-1}, generally determined up to an additive constant, is given for $f \in \mathcal{F}$ by

$$\theta^{-1} f(z) = \int_0^z f(\zeta)\, d \log \zeta$$

$$= \int_0^z \frac{f(\zeta)}{\zeta} d\zeta.$$

Hence, we see that $\mathcal{L} f(1)(z) = \theta^{-1} f(z)$. The generalization is defined by $\mathcal{D}$:

$$\mathcal{D}^{-\lambda}\psi(w) = \frac{1}{\Gamma(\lambda)} \int_{\infty}^{w} \psi(\omega)(w - \omega)^{\lambda-1} d\omega, \tag{6.3}$$

where the integration is taken along the half-straight line parallel to the real axis which is contained in the left half-plane or along any equivalent path.

If we put $w = \log z$ and accordingly $\omega = \log \zeta$, then we have $\mathcal{D}^{-1} = \theta^{-1}$ where the functions to be applied are of course connected by $\psi(w) = f(z)$ with $w = \log z$. The relation between $\mathcal{D}^{-\lambda}$ and $\theta^{-\lambda}$ is consequently given by

$$\mathcal{D}^{-\lambda} = \theta^{-\lambda} \qquad (w = \log z). \tag{6.4}$$

THEOREM 6.2. *The operator* $\mathcal{L}(1)^{\lambda}$ *coincides with* $\theta^{-\lambda}$, *that is, for any* $f \in \mathcal{F}$ *we have*

$$\mathcal{L}(1)^{\lambda} f(z) = \theta^{-\lambda} f(z).$$

Proof. By putting $w = \log z$ and accordingly $\omega = \log \zeta$, the complex integral form (6.2) becomes

$$\mathcal{L}(1)^{\lambda} f(z) = \frac{1}{\Gamma(\lambda)} \int_{0}^{w} f(e^{w})(w - \omega)^{\lambda-1} d\omega.$$

This shows just

$$\mathcal{L}(1)^{\lambda} f(z) = \mathcal{D}^{-\lambda} f(e^{w}) = \theta^{-\lambda} f(z). \qquad \square$$

By means of the relation shown in Theorem 6.2, we can establish that the inversion formula for

$$\mathcal{L}(1)^{\lambda} f(z) = f_{\lambda}(z)$$

is given, as shown in the following theorem:

THEOREM 6.3. *For any* $\lambda > 0$, *put* $p = -[-\lambda]$ *and* $\delta = p - \lambda$, $[\ ]$ *denoting Gauss' symbol. Then the inversion formula for* $\mathcal{L}^{\lambda}(1) f(z) = f_{\lambda}(z)$ *is given by*

$$f(z) = \theta^{p} \mathcal{L}(1)^{\delta} f_{\lambda}(z),$$

that is,

$$f(z) = \left(z\frac{d}{dz}\right)^{p} \frac{1}{\Gamma(\lambda)} \int_0^z \frac{f_{\lambda}(\zeta)}{\zeta} \left(\log\frac{z}{\zeta}\right)^{\delta-1} d\zeta.$$

Proof. Since the inverse of $\theta^{-\lambda}$ is given by θ^{λ}, we obtain in view of Theorem 6.2, from $f_{\lambda}(z) = \mathcal{L}^{\lambda} f(z) = \theta^{-\lambda} f(z)$ the relation

$$f(z) = \theta^{\lambda} f_{\lambda}(z) = \theta^{p} \theta^{-\delta} f_{\lambda}(z) = \theta^{p} \mathcal{L}(1)^{\delta} f_{\lambda}(z),$$

which is the desired one. □

Once the inversion formula having been established, it is desirable to extend the domain of definition of $\mathcal{L}(1)^{\lambda}$ into the negative value of λ. In fact, the relation in Theorem 6.2 has been shown for $\lambda > 0$ but its right-hand member has a definite meaning also as the fractional derivative of order $-\lambda$. Accordingly, we may regard this relation itself as the defining equation of $\mathcal{L}(1)^{\lambda}$ with $\lambda < 0$. Thus, for $\lambda < 0$, we put $\mathcal{L}(1)^{\lambda} = (\mathcal{L}(1)^{-\lambda})^{-1}$, that is, $\mathcal{L}(1)^{\lambda} = \theta^{-\lambda}$ or

$$\mathcal{L}(1)^{\lambda} f(z) = \left(z\frac{d}{dz}\right)^{p} \frac{1}{\Gamma(\delta)} \int_0^z \frac{f(\zeta)}{\zeta} \left(\log\frac{z}{\zeta}\right)^{\delta-1} d\zeta,$$

where $p = -[\lambda]$ and $\delta = p + \lambda$. This definition agrees essentially with the relation given in Theorem 6.3. The family of operators $\{\mathcal{L}(1)^{\lambda}\}_{-\infty<\lambda<+\infty}$ then becomes a commutative group with respect to composition.

THEOREM 6.4. *For any λ, positive or negative, the operator $\mathcal{L}(1)^{\lambda}$ is represented by a unified expression in terms of series expansion*

$$\mathcal{L}(1)^{\lambda} f(z) = \sum_{\nu=1}^{\infty} \frac{c_{\nu}}{\nu^{\lambda}} z^{\nu}$$

where

$$f(z) = \sum_{\nu=1}^{\infty} c_{\nu} z^{\nu} \in \mathcal{F}.$$

The additivity condition

$$\mathcal{L}(1)^{\lambda} \mathcal{L}(1)^{\mu} = \mathcal{L}(1)^{\lambda+\mu}$$

remains to hold for any pair of values λ and μ.

Proof. Since the case of non-negative λ has been already known, here we only have to verify the series expansion for $\lambda < 0$. By means of its definition we get really

$$\mathcal{L}(1)^{\lambda} f(z) = \left(z \frac{d}{dz}\right)^{p} \mathcal{L}(1)^{\delta} f(z)$$

$$= \left(z \frac{d}{dz}\right)^{p} \sum_{\nu=1}^{\infty} \frac{c_{\nu}}{\nu^{\delta}} z^{\nu}$$

$$= \sum_{\nu=1}^{\infty} \frac{c_{\nu}}{\nu^{\delta-p}} z^{\nu}$$

$$= \sum_{\nu=1}^{\infty} \frac{c_{\nu}}{\nu^{\lambda}} z^{\nu}.$$

The assertion on additivity is evident from the series expansion. □

REMARK. Since the sequence $\{1/\nu^{\nu}\}_{\nu}$ is increasing for $\lambda < 0$, the operator $z^{-\lambda}\mathcal{L}(1)^{\lambda}$ with $\lambda < 0$ is regarded as a particular so-called Gel'fond-Leont'ev derivative introduced by themselves [1].

§ 7. Relations to integration operator

In the previous section we have dealt with a special probability measure $\sigma(t) = \sigma(t\,;\,1) = t$ which generates the family of operators $\{\mathcal{L}(1)^{\lambda}\}$. This family is represented by the complex form:

$$\mathcal{L}(1)^{\lambda} f(z) = \frac{1}{\Gamma(\lambda)} \int_I \frac{f(zt)}{t} \left(\log \frac{1}{t}\right)^{\lambda-1} dt$$

$$\left[\, t = \frac{\zeta}{z} \,\right]$$

$$= \frac{1}{\Gamma(\lambda)} \int_{\infty}^{\log z} f(\zeta)(\log z - \log \zeta)^{\lambda-1} d\log \zeta.$$

Here, in the last integration, the path is taken along the half straight line on the $\log \zeta$-plane which is parallel to the real axis and contained in the left half-plane $\{\mathrm{Re}\,\log \zeta < 0\}$. Thus, this operator coincides with the fractional integration of order λ with respect to $\log z$. In particular, $\mathcal{L}(1)$ is just the inverse operator of

$$\theta = \frac{d}{d\log z}.$$

The last-mentioned fact is peculiar to the case $a = 1$. The corresponding property in the case $a \neq 1$ is stated as in the following Theorem; cf. Komatu [14].

THEOREM 7.1. *The operator* $\mathcal{L}(a)$ *with* $a \neq 1$ *coincides with the integration with respect to* $w = a(a-1)^{-1} z^{a-1}$ *followed by multiplication of* $z^{-(a-1)}$, *the non-integral powers being understood to mean the principal branch. More specifically, we have*

$$\mathcal{L}(a) f(z) = \begin{cases} \dfrac{1}{z^{a-1}} \displaystyle\int_{\infty}^{w} F(\varphi)\, d\varphi & (0 < a < 1), \\[2ex] \dfrac{1}{z^{a-1}} \displaystyle\int_{0}^{w} F(\varphi)\, d\varphi & (a > 1), \end{cases}$$

where

$$w = \frac{a}{a-1} z^{a-1}, \qquad F(z) = f\left(\frac{a-1}{a} w\right)^{1/(a-1)}$$

and the integration paths are the half straight line

$\{\arg \varphi = \pi - (1-a)\arg z,\ \infty > |\varphi| > |w|\}$ *for* $0 < a < 1$

and the segment

$\{\arg \varphi = (a-1)\arg z,\ 0 < |\varphi| < |w|\}$ *for* $a > 1$,

respectively.

Proof. The operator $\mathcal{L}(a)$ is, by definition, given by

$$\mathcal{L}(a)f(z) = a\int_I \frac{f(zt)}{t}\, t^{a-1}\, dt$$

$$= \frac{a}{z^{a-1}}\int_0^z f(\zeta)\zeta^{a-2}\, d\zeta,$$

the last integration being taken along the segment from 0 to z. We only have to change the integration variable of the last integral by

$$d\varphi = a\zeta^{a-2}\, d\zeta,$$

or more concretely, by

$$\varphi = a(a-1)^{-1}\zeta^{a-1}.$$

When ζ runs along the segment from 0 to z, φ runs along just the respective integration paths stated in the Theorem. □

It will be seen that the relation

$$\frac{d}{d\log z}\mathcal{L}(a)f(z) = a f(z) - (a-1)\mathcal{L}(a)f(z)$$

holds for any $a > 0$. This may be regarded as a straightforward generalization of the already mentioned relation

$$\frac{d}{d\log z}\mathcal{L}(1)f(z) = f(z)$$

corresponding to the case $a = 1$. Here we state it in slightly general form.

THEOREM 7.2. *For any* $a > 0$ *and* $\lambda \geq 1$, *we have*

$$\theta \mathcal{L}(a)^{\lambda} = a \mathcal{L}(a)^{\lambda-1} - (a-1)\mathcal{L}(a)^{\lambda},$$

$\mathcal{L}(a)^{0}$ *being understood to be the identity operator.*

Proof. By differentiating logarithmically the defining equation of $\mathcal{L}(a)^{\lambda} f(z)$, we obtain

$$\frac{d}{d\log z} \mathcal{L}(a)^{\lambda} f(z)$$

$$= \frac{a^{\lambda}}{\Gamma(\lambda)} \int_I z f'(zt)\, t^{a-1}\left(\log \frac{1}{t}\right)^{\lambda-1} dt$$

which becomes after integration by parts

$$\frac{d}{d\log z} \mathcal{L}(a)^{\lambda} f(z)$$

$$= \frac{a^{\lambda}}{\Gamma(\lambda)} \left[f(zt)\, t^{a-1}\left(\log \frac{1}{t}\right)^{\lambda-1} \right]_0^1$$

$$- \int_I f(zt)\left(-(\lambda-1)\, t^{a-2}\left(\log \frac{1}{t}\right)^{\lambda-2}\right.$$

$$\left. + (a-1)\, t^{a-2}\left(\log \frac{1}{t}\right)^{\lambda-1}\right) dt .$$

Here we remember $f \in \mathcal{F}$ and $a > 0$. We get for $\lambda = 1$

$$\frac{d}{d\log z}\mathcal{L}(a)f(z)$$

$$= a\left(f(z) - (a-1)\int_I f(zt)t^{a-2}\,dt\right)$$

$$= a\,(f(z) - (a-1)\mathcal{L}(a)f(z)),$$

while we obtain for $\lambda > 1$

$$\frac{d}{d\log z}\mathcal{L}(a)^{\lambda}f(z)$$

$$= \frac{a^{\lambda}}{\Gamma(\lambda)}\left((\lambda-1)\int_I f(zt)t^{a-2}\left(\log\frac{1}{t}\right)^{\lambda-2}dt\right.$$

$$\left. - (a-1)\int_I f(zt)t^{a-2}\left(\log\frac{1}{t}\right)^{\lambda-1}dt\right)$$

$$= a\mathcal{L}(a)^{\lambda-1}f(z) - (a-1)\mathcal{L}(a)^{\lambda}f(z). \qquad \square$$

In the following lines, we shall consider the relation of $\mathcal{L}(a)$ to the ordinary integration operator $\mathcal{J}$ defined by

$$\mathcal{J}f(z) = \int_0^z f(\zeta)\,d\zeta.$$

For that purpose, we attempt to derive an expression for $\mathcal{L}(a)$ in terms of $\mathcal{J}$ and its iterations.

It is well-known that its iteration is represented by the following form:

$$\mathcal{J}^{\kappa} f(z) = \int_0^z \mathcal{J}^{\kappa-1} f(\zeta)\, d\zeta$$

$$= \int_0^z \int_0^\zeta \cdots \int_0^\zeta f(\zeta)(d\zeta)^{\kappa-1}$$

$$= \frac{1}{(\kappa - 1)!} \int_0^z f(\zeta)(z - \zeta)^{\kappa-1} d\zeta.$$

On the other hand, for the sake of brevity, we make use of Pochhammer's symbol, which is defined by

$$(x)_n = \frac{\Gamma(x + n)}{\Gamma(x)}$$

$$= \prod_{j=0}^{n-1} (x + j) \qquad (n = 0, 1, \ldots),$$

the empty product denoting unity; in particular,

$$(x)_0 = 1$$

even for $x = 0$.

Now, we shall mention an interrelation between $\mathcal{L}(a)$ and $\{\mathcal{J}^{\kappa}\}_{\kappa}$:

THEOREM 7.3. *For any* $a > 0$ *we have*

$$\mathcal{L}(a) = a\sum_{\kappa=1}^{\infty} \frac{(2-a)_{\kappa-1}}{z^{\kappa}} \mathcal{J}^{\kappa}.$$

In particular, when $a = k > 1$ is an integer, the right hand expression reduces to a finite sum consisting of the beginning $k - 1$ terms.

Proof. Since $|z - \zeta| < |z|$ holds on the integration path in the expression for $\mathcal{L}(a)$ except at $\zeta = 0$, we have

$$\zeta^{a-2} = z^{a-2}\left(1 - \frac{z-\zeta}{z}\right)^{a-2}$$

$$= z^{a-2}\sum_{\kappa=0}^{\infty} (-1)^{\kappa} \binom{a-2}{\kappa}\left(\frac{z-\zeta}{z}\right)^{\kappa}$$

$$= z^{a-2}\sum_{\kappa=1}^{\infty} \frac{(2-a)_{\kappa-1}}{(\kappa-1)!} \frac{1}{z^{\kappa-1}} (z-\zeta)^{\kappa-1}.$$

Substitution followed by termwise integration yields

$$\mathcal{L}(a) f(z)$$

$$= \frac{a}{a-1}\int_0^z f(\zeta)\zeta^{a-2}\, d\zeta$$

$$= a\sum_{\kappa=1}^{\infty} \frac{(2-a)_{\kappa-1}}{z^{\kappa}} \frac{1}{(\kappa-1)!}\int_0^z f(\zeta)(z-\zeta)^{\kappa-1}\, d\zeta$$

$$= a \sum_{\kappa=1}^{\infty} \frac{(2-a)_{\kappa-1}}{z^{\kappa}} \mathcal{J}^{\kappa} f(z).$$

When $a = k > 1$ is an integer, then $(2-k)_{\kappa-1}$ vanishes for every $\kappa \geq k$. □

REMARK. It may be noted that, for integral value of $a > 1$, the relation in Theorem 7.3 can also be inductively verified, by making use of integration by parts.

On the other hand, the case $a = 1$ in the Theorem is exceptional in the sense that every term in the summand for $\mathcal{L}(1)$ does not vanish.

§ 8. Generalizations

Now, we conider a new class of functions $\{F\}$, which consists of functions F defined by

$$F(z) := P[\theta] f(z),$$

where P and θ are given by

$$P[\theta] = \sum_{\kappa=0}^{K} A_{\kappa} \theta^{\kappa} \quad \text{and} \quad \theta = \frac{d}{d \log z};$$

cf. Komatu [19].

LEMMA 8.1. *If $f \in \mathcal{J}$, then any function $F(z) = P[\theta] f(z)$ or equivalently*

$$F(z) = \sum_{\kappa=0}^{K} B_\kappa z^\kappa \left(\frac{d}{dz}\right)^\kappa f(z)$$

with any positive integer K belongs to $\mathcal{F}$, provided the coefficients satisfy a condition, which is independent of the choice of $f \in \mathcal{F}$,

$$\sum_{\kappa=0}^{K} A_\kappa = 1 \quad \text{or} \quad B_0 + B_1 = 1.$$

Proof. Recall the normalization on $f \in \mathcal{F}$ that

$$f(z) = \sum_{\nu=1}^{\infty} c_\nu z^\nu, \quad c_1 = 1.$$

Then $F = P[\theta] f$ has the expansion

$$F(z) = \sum_{\kappa=0}^{K} A_\kappa \sum_{\nu=1}^{\infty} c_\nu z^\nu, \quad c_1 = 1.$$

In particular, $F(0) = 0$ and $F'(0) = A_0 + \ldots + A_K = B_0 + B_1$. Thus the conclusion is clear. □

By comparing the acts mentioned in the Lemma 13.2 just proved, which cause F from f, we have

$$\sum_{\kappa=0}^{K} A_\kappa \left(\frac{d}{d \log z}\right)^\kappa = \sum_{\kappa=0}^{K} B_\kappa z^\kappa \left(\frac{d}{dz}\right)^\kappa .$$

It follows that every B_κ with $\kappa \geqq 0$ is a linear combination of $\{A_j\}_{j=\kappa}^{K}$ with integral coefficients of positive sign. In fact, the coefficients in the equality

$$\left(\frac{d}{d\log z}\right)^j = \sum_{\kappa=1}^{j} \alpha_{j\kappa}\, z^{\kappa} \left(\frac{d}{dz}\right)^{\kappa}$$

are determined by recurrence relations

$$\alpha_{11}=1, \quad \alpha_{j\kappa} = \alpha_{j-1,\kappa-1} + \kappa\,\alpha_{j-1,\kappa} \qquad (1 \leqq \kappa \leqq j)$$

with convention $\alpha_{j-1,0} = \alpha_{j-1,j} = 0$. In particular, it is seen that $\alpha_{\kappa\kappa} = 1$. Hence, we have

$$\sum_{j=1}^{K} A_j \left(\frac{d}{d\log z}\right)^j = \sum_{j=1}^{K} A_j \sum_{\kappa=1}^{j} \alpha_{j\kappa}\, z^{\kappa} \left(\frac{d}{dz}\right)^{\kappa}$$

$$= \sum_{\kappa=1}^{K} \left(\sum_{j=\kappa}^{K} \alpha_{j\kappa} A_j\right) \left(\frac{d}{dz}\right)^{\kappa},$$

whence follows

$$B_{\kappa} = \sum_{j=\kappa}^{K} \alpha_{j\kappa} A_j \qquad (0 \leq \kappa \leq K)$$

with $\alpha_{\kappa\kappa} = 1$ and, in particular,

$$B_1 = \sum_{j=1}^{K} A_j, \qquad B_K = A_K.$$

Consequently, every A_κ with $\kappa \geqq 0$ is conversely a linear combination of $\{B_j\}_{j=\kappa}^{K}$ with integral coefficients of indefinite sign; and, in particular,

$$F'(0) = B_0 + B_1 = \sum_{\kappa=0}^{K} A_\kappa = 1.$$

Finally, we insert here a simple example for illustration. If we take $K = 2$, the generic form of $F \in \mathcal{F}$ is given either of the forms

$$F(z) = \left(A_0 + (1 - A_0 - A_2)\frac{d}{d\log z} + A_2\left(\frac{d}{d\log z}\right)^2 \right) f(z),$$

$$F(z) = A_0 f(z) + (1 - A_0)zf'(z) + A_2 z^2 f''(z)$$

with arbitrary constants A_0 and A_2. Hence we have

$$\mathcal{L}F(z) = \left(A_0 + (1 - A_0 - A_2)\frac{d}{d\log z} + A_2\left(\frac{d}{d\log z}\right)^2 \right)\mathcal{L}f(z).$$

If we restrict ourselves to the distinguished case generated by $\sigma(t) = t$, then the corresponding $\mathcal{L}$ becomes the integration with respect to $\log z$, so that we have

$$\mathcal{L}F(z) = A_0 \int_0^z \frac{f(\zeta)}{\zeta}\, d\zeta + (1 - A_0 - A_2)f(z) + A_2 z f'(z).$$

As an illustrative purpose, we mention here a remark. In fact, if we put $A_0 = a_0 + ib_0$ and $A_2 = a_2 + ib_2$ with real a_0, b_0, a_2 and b_2, then we have

$$\operatorname{Re} \frac{F(z)}{z} = a_0 \operatorname{Re} \frac{f(z)}{z} + (1 - a_0)\operatorname{Re} f'(z) + a_2 \operatorname{Re} zf''(z) - b_0 \operatorname{Im} \frac{f(z)}{z} + b_0 \operatorname{Im} f'(z) - b_2 \operatorname{Im} zf''(z),$$

$$\operatorname{Re} \frac{\mathcal{L}F(z)}{z} = a_0 \operatorname{Re} \frac{1}{z} \int_0^z \frac{f(\zeta)}{\zeta}\, d\zeta + (1 - a_0 - a_2) \operatorname{Re} \frac{f(z)}{z} + a_2 \operatorname{Re} f'(z) - b_0 \operatorname{Im} \frac{1}{z} \int_0^z \frac{f(\zeta)}{\zeta}\, d\zeta + (b_0 + b_2) \operatorname{Im} \frac{f(z)}{z} - b_2 \operatorname{Im} f'(z).$$

If, in particular, $A_0 = 0$ and $A_2 = -i$, then

$$\operatorname{Re} \frac{F(z)}{z} = \operatorname{Re} f'(z) + \operatorname{Im} zf''(z),$$

$$\operatorname{Re} \frac{\mathcal{L}F(z)}{z} = \operatorname{Re} \frac{f(z)}{z} - \operatorname{Im} \frac{f(z)}{z} + \operatorname{Im} f'(z).$$

Next, by relaxing the restriction that the referring probability measure σ is a monomial, we now consider a probability measure defined by means of power series

$$\sigma(t) = \sum_{k=1}^{\infty} \omega_k t^k \tag{8.1}$$

with convergence radius greater than unity:

$$\limsup_{k \to \zeta} |\omega_k|^{1/k} < 1;$$

cf. Komatu [14].

In view of the condition that σ is a probability measure, we have to suppose

$$\rho(t) \equiv \sigma'(t)$$

$$\begin{aligned} &= \sum_{k=1}^{\infty} k\omega_k t^{k-1} \geqq 0 \quad (t \in I); \\ \sigma(1) &= \sum_{k=1}^{\infty} \omega_k = 1. \end{aligned} \tag{8.2}$$

The operator generated by (8.1) satisfying (8.2) will be denoted by $\mathcal{L}[\rho]$. We begin with the following theorem:

THEOREM 8.1. *Let* σ *satisfy the conditions* (8.1) *and* (8.2) *above. Then, the operator* $\mathcal{L}[\rho]$ *defined by*

$$\mathcal{L}[\rho]f(z) = \int_I \frac{f(zt)}{t} \rho(t)\,dt \quad (f \in \mathcal{F})$$

is represented in terms of the ordinary integration operator $\mathcal{J}$ *in the form*

$$\mathcal{L}[\rho] = \sum_{\kappa=1}^{\infty} (-1)^{\kappa-1} \frac{\varphi^{(\kappa-1)}(1)}{z^{\kappa}} \mathcal{J}^{\kappa}.$$

where φ *is defined by*

$$\varphi(t) = \frac{\rho(t)}{t} = \sum_{k=1}^{\infty} k\,\omega_k\, t^{k-2}.$$

Proof. In view of (8.1) and (8.2) we have

$$\mathcal{L}[\rho] = \sum_{k=1}^{\infty} \omega_k \mathcal{L}(k). \tag{8.3}$$

By substituting the expressions for $\mathcal{L}(k)$ $(k = 1, 2, \ldots)$ derived in Theorem 7.3 into (8.3), we obtain

$$\begin{aligned} \mathcal{L}[\rho] &= \sum_{k=1}^{\infty} \omega_k \mathcal{L}(k) \\ &= \omega_1 \sum_{\kappa=1}^{\infty} \frac{(\kappa-1)!}{z^{\kappa}} \mathcal{J}^{\kappa} \\ &\quad + \sum_{k=2}^{\infty} \omega_k\, k \sum_{\kappa=1}^{k-1} (-1)^{\kappa-1} \frac{(k-2)!}{(k-\kappa-1)!\, z^{\kappa}} \mathcal{J}^{\kappa} \\ &= \sum_{\kappa=1}^{\infty} \frac{\phi_{\kappa}}{z^{\kappa}} \mathcal{J}^{\kappa}, \end{aligned}$$

where the coefficients of the last expression are given by

$$\phi_\kappa = (\kappa - 1)!\ \omega_1 + (-1)^{\kappa-1} \sum_{k=\kappa+1}^{\infty} k \frac{(k-2)!}{(k-\kappa-1)!} \omega_k$$

$$= (-1)^{\kappa-1} \left[\frac{d^{\kappa-1}}{dt^{\kappa-1}} \left(\frac{\omega_1}{t} + \sum_{k=2}^{\infty} k\,\omega_k\, t^{k-2} \right) \right]^{t=1}$$

$$= (-1)^{\kappa-1}\ \varphi^{(\kappa-1)}(1).$$

We thus get the desired result. □

We have considered the basic measure given by (8.1) and derived Theorem 8.1. This can be slightly further generalized with respect to the referring measure. That is, we have the the following result:

THEOREM 8.2. *Let a probability measure* σ *be given by*

$$\sigma(t) = \int_0^\infty t^a\, d\tau(a)$$

where a measure τ *defined on the interval* $(0, \infty)$ *satisfies the conditions*

$$\rho(t) \equiv \sigma'(t)$$

$$= \int_0^\infty a\, t^{a-1}\, d\tau(a) \geqq 0 \qquad (t \in I),$$

$$\sigma(1) = \int_0^\infty d\tau(a) = 1.$$

Then, we have

$$\mathcal{L}[\rho] = \sum_{\kappa=1}^{\infty} (-1)^{\kappa-1} \frac{\varphi^{(\kappa-1)}(1)}{z^{\kappa}} \mathcal{J}^{\kappa},$$

where φ *is defined by*

$$\varphi(t) = \frac{\rho(t)}{t} = \int_0^{\infty} a\, t^{a-2}\, d\tau(a).$$

Proof. The proof proceeds quite similarly to that of the previous theorem. We have

$$\mathcal{L}[\rho] f(z) = \int_I \frac{f(zt)}{t}\, d\int_0^{\infty} t^{a}\, d\tau(a)$$

$$= \int_0^{\infty} d\tau(a) \int_I \frac{f(zt)}{t}\, at^{a-1} dt$$

$$= \int_0^{\infty} \mathcal{L}(a) f(z)\, d\tau(a).$$

By substituting the expression for $\mathcal{L}(a)$ derived in Theorem 7.3, we obtain

$$\mathcal{L}[\rho] = \int_0^{\infty} a \sum_{\kappa=1}^{\infty} \frac{(2-a)_{\kappa-1}}{z^{\kappa}} \mathcal{J}^{\kappa}\, d\tau(a)$$

$$= \sum_{\kappa=1}^{\infty} \frac{\phi_{\kappa}}{z^{\kappa}} \mathcal{J}^{\kappa},$$

where the coefficients of the last expression are given by

$$\phi_\kappa = \int_0^\infty a\,(2-a)_{\kappa-1}\, d\tau(a)$$

$$= (-1)^{\kappa-1}\left[\frac{d^{\kappa-1}}{dt^{\kappa-1}}\int_0^\infty at^{a-2}\, d\tau(a)\right]^{t=1}$$

$$= (-1)^{\kappa-1}\varphi^{(\kappa-1)}(1).$$

Thus we get the desired result. □

REMARK. Throughout this section the restriction

$$\rho(t) = \sigma'(t) \geqq 0 \qquad (t \in I)$$

is really inessential, since the whole discussion appeared has concerned exhaustively to derive relations involving e-quality. From this standpoint, we shall supplement in the next section some examples concerning Theorem 8.2.

§ 9. Examples

In the present section, in particular, in relation with the Remark mentioned at the end of the preceding section, we shall illustrate some examples concerning theorems discussed until now; cf. Komatu [14].

Example 1. Let χ be the unit function with respect to Hadamard product:

$$\chi(z) = \frac{z}{1 - z} = \sum_{\nu=1}^{\infty} z^{\nu}.$$

Noting that its expansion coefficients are all equal to unity and that the moment sequence with respect to $\mathcal{L}(a)$ is

$$\alpha_{\nu}(a) = \frac{a}{a + \nu - 1} \quad (\nu = 1, 2, \ldots),$$

we have

$$\mathcal{L}(a)\,\chi(z) = a \sum_{\nu=1}^{\infty} \frac{z^{\nu}}{a + \nu - 1}.$$

On the other hand, we have derived in Theorem 7.3 an expression for $\mathcal{L}(a)$ in term of $\{\mathcal{J}^{\kappa}\}$; this in particular yields

$$\mathcal{L}(a)\,\chi(z) = a \sum_{\kappa=1}^{\infty} \frac{(2 - a)_{\kappa-1}}{z^{\kappa}}\, \mathcal{J}^{\kappa}\,\chi(z).$$

By comparing these expressions, we obtain the relation

$$\sum_{\lambda=1}^{\infty} \frac{z^{\nu}}{a + \nu - 1} = \sum_{\kappa=1}^{\infty} \frac{(2 - a)_{\kappa-1}}{z^{\kappa}}\, \mathcal{J}^{\kappa}\,\chi(z)$$

valid for $a > 0$. It is readily seen directly that $\mathcal{J}^{\kappa}$ is expressed by the expansion

$$\mathcal{J}^{\kappa}\,\chi(z) = z^{\kappa} \sum_{\nu=1}^{\infty} \frac{z^{\nu}}{(\nu + 1)_{\kappa}}.$$

By substituting this into the above relation and comparing the coefficients of z^{ν} in both sides, we obtain the identity

$$\frac{1}{a+\nu-1} = \sum_{\kappa=1}^{\infty} \frac{(2-a)_{\kappa-1}}{(\nu+1)_{\kappa}} \qquad (\nu = 1, 2, \ldots).$$

We notice here that $\mathcal{J}^{\kappa}\chi(z)$ is for any integer $\kappa \geq 0$ an elementary function of z. We thushave, for instance,

$$\mathcal{J}\chi(z) = \log\frac{1}{1-z} - z,$$

$$\mathcal{J}^{2}\chi(z) = -(1-z)\log\frac{1}{1-z} + z - \frac{z^{2}}{2}.$$

For any integer $\kappa \geqq 2$ we can derive similar explicit expression in the form

$$\begin{aligned}
&\mathcal{J}^{\kappa}\chi(z)\\
&= \frac{(-1)^{\kappa-1}}{(\kappa-1)!}(1-z)^{\kappa-1}\log\frac{1}{1-z} + \frac{(-1)^{\kappa-1}}{\kappa!}(1-z)^{\kappa}\\
&+ \frac{(-1)^{\kappa-1}}{(\kappa-1)!}\sum_{j=2}^{\kappa-1}\frac{1}{j}\,(1-z)^{\kappa-1}\\
&+ \sum_{j=0}^{\kappa-2}\frac{(-1)^{j}}{j!}\,\frac{1}{(\kappa-j)!(\kappa-j-1)}(1-z)^{j},
\end{aligned}$$

the empty sum being understood zero. The last relation is verified, for instance, by induction, where the actual calculation is somewhat lengthy.

Example 2. (i) Let us consider the probability measure given by

$$\sigma(t) = \frac{e^{t} - 1}{e - 1}.$$

It has the density

$$\rho(t) = \sigma'(t) = \frac{e^{t}}{e - 1},$$

and hence, by means of Theorem 8.2, we obtain

$$\mathcal{L}[\rho] = \sum_{\kappa=1}^{\infty} (-1)^{\kappa-1} \frac{\varphi^{(\kappa-1)}(1)}{z^{\kappa}} \mathcal{J}^{\kappa},$$

where φ is defined by

$$\varphi(t) = \frac{\rho(t)}{t} = \frac{1}{e - 1} \frac{e^{t}}{t}.$$

(ii) Let us consider the probability measure given by

$$\sigma(t) = \frac{1 - e^{-t}}{1 + e^{-1}}.$$

In this case, discussions proceed quite similarly as above. Namely, it has the density

$$\rho(t) = \sigma'(t) = \frac{e^{-t}}{1 - e^{-1}}.$$

Hence, by means of Theorem 8.2, we obtain

$$\mathcal{L}[\rho] = \sum_{\kappa=1}^{\infty} (-1)^{\kappa-1} \frac{\varphi^{(\kappa-1)}}{z^{\kappa}} \mathcal{J}^{\kappa},$$

where φ is defined by

$$\varphi(t) = \frac{\rho(t)}{t} = \frac{1}{1-e^{-1}} \frac{e^{-t}}{t}.$$

Example 3. According to the remark stated at the end of preceding section, we consider σ' (of indefinite sign) given by

$$\sigma(t) = A_{2m} \int_0^t \tau P_{2m}(\tau)\, d\tau \qquad (m \geqq 1),$$

where P_{2m} denotes the Legendre polynomial of degree $2m$ and A_{2m} is the normalization factor determined by $\sigma(1) = 1$.

By means of the Rodrigues formula, we get after repeated integration by parts

$$\frac{1}{A_{2m}} = \int_I \tau P_{2m}(\tau)\, d\tau$$

$$= \frac{(-1)^{m-1}}{2^{2m}} \frac{(2m-2)!}{(m-1)!(m+1)!}.$$

By making use of a familiar formula

$$P_n(t) = \sum_{\nu=0}^{n} (-1)^\nu \frac{(n+\nu)!}{\nu!^2(n-\nu)!}\left(\frac{1-t}{2}\right)^\nu,$$

we get after repeated differentiation

$$P_n^{(\nu-1)}(t)$$

$$= \sum_{\nu=\kappa-1}^{n} (-1)^\nu \frac{(n+\nu)!}{\nu!^2(n-\nu)!}\left(-\frac{1}{2}\right)^{\kappa-1} \frac{\nu!}{(\nu-\kappa+1)!}\left(\frac{1-t}{2}\right)^{\nu-\kappa+1}.$$

Setting

$$\varphi(t) = A_{2m} P_{2m}(t),$$

we obtain the value of $\varphi^{(\kappa-1)}$ and finally

$$\mathcal{L}[\sigma'] = (-1)^{m-1} 2^m \frac{(m-1)!(m+1)!}{(2m-2)!}$$

$$\sum_{\kappa=1}^{2m+2} \frac{(-1)^{\kappa-1}(2m+\kappa-1)!}{(\kappa-1)!(2m-\kappa+1)!} \frac{1}{2^{\kappa-1}} \frac{1}{z^\kappa} \varphi^\kappa .$$

For $m = 0$, we have

$$P_0(\tau) = 1, \ A_0 = 2; \quad \sigma(t) = t^2,$$

$$\mathcal{L}[\sigma'] = \mathcal{L}(2) = \frac{2}{z} \mathcal{J} .$$

Even for $m = 1/2$,

$$P_1(\tau) = \tau, \; A_1 = 3; \qquad \sigma(t) = t^3,$$

$$\mathcal{L}[\sigma'] = \mathcal{L}(3) = 3\frac{1}{z}\mathcal{J} - \frac{1}{z^2}\mathcal{J}^2.$$

However, the case of odd $2m > 1$ has been excluded, since we would have $1/A_n = 0$ for any odd integer $n \geq 3$.

Chapter 3. Applications

§ 10. Product of operators

As we know, the family of operators $\{\mathcal{L}(a)^{\lambda}\}_{\lambda\geqq 0}$ defined by

$$\mathcal{L}(a)^{\lambda} f(z) = \frac{a^{\lambda}}{\Gamma(\lambda)} \int_I f(zt)\, t^{a-2} \left(\log \frac{1}{t}\right)^{\lambda-1} dt$$

which depends on a continuous parameter $a \geqq 0$ possesses the semiring structure:

$$\mathcal{L}(a)^{\lambda} \mathcal{L}(a)^{\mu} = \mathcal{L}(a)^{\lambda+\mu}.$$

In connection with this additivity relation, we shall take a step forward and consider in the following lines the product of operators of such type with several different a's; cf. Komatu [16]. For simplicity, we restrict ourselves to the positive integral values of λ's.

To simplify the description, we set

$$(10.1) \qquad \mathcal{K}(a)^{h} = \frac{\mathcal{L}(a)^{h}}{a^{h}} \qquad (h = 1, 2, \ldots)$$

in the following lines; in particular,

$$\mathcal{L} = \mathcal{L}^1 \quad \text{and} \quad \mathcal{K} = \mathcal{K}^1.$$

We begin with a lemma, which shows a resolvent equation; cf. Theorem 11.3 below:

LEMMA 10.1. *We have*

$$\mathcal{K}(a)\mathcal{K}(b) = \begin{cases} \dfrac{1}{b-a}(\mathcal{K}(a) - \mathcal{K}(b)) & (b \neq a), \\ -\dfrac{d}{da}\mathcal{K}(a) & (b = a). \end{cases}$$

Proof. In view of definition of $\mathcal{L}$, we get

$$\begin{aligned} &\mathcal{L}(a)\mathcal{L}(b)f(z) \\ &= a\int_I s^{a-2}\, ds\, b\int_I f(zs\tau)\tau^{b-2}\, d\tau \\ &= ab\int_I s^{a-2}\, ds \int_0^s f(zt)\left(\frac{t}{s}\right)^{b-2}\frac{dt}{s} \\ &= ab\int_I f(zt)t^{b-2}\, dt \int_t^1 s^{a-b-1}\, ds \\ &= \int_I \frac{f(zt)}{t}\rho(t)\, dt, \end{aligned}$$

where the density ρ is given by

$$\rho(t) = abt^{b-1}\int_t^1 s^{a-b-1}\, ds$$

$$= \begin{cases} \dfrac{ab}{a-b}\left(t^{b-1} - t^{a-1}\right) & (b \neq a), \\ \\ a^2 t^{a-1} \log \dfrac{1}{t} & (b = a). \end{cases}$$

Hence we obtain, if $b \neq a$

$$\mathcal{L}(a)\,\mathcal{L}(b) = \frac{1}{a-b}\left(a\,\mathcal{L}(b) - b\,\mathcal{L}(a)\right),$$

while, if $b = a$, by taking the definition of $\mathcal{L}(a)^2$ into account,

$$\mathcal{L}(a)^2 f(z) = \int_I \frac{f(zt)}{t}\, a^2\, t^{a-1} \log \frac{1}{t}\, dt$$

$$= \int_I \frac{f(zt)}{t}\left(at^{a-1} - a\frac{d}{da}\left(at^{a-1}\right)\right) dt$$

$$= \left(\mathcal{L}(a) - a\frac{d}{da}\mathcal{L}(a)\right) f(z)$$

$$= -a^2 \frac{d}{da}\frac{\mathcal{L}(a)}{a} f(z).$$

Writing these expressions for $\mathcal{L}(a)\,\mathcal{L}(b)$ and $\mathcal{L}(a)^2$ in terms of $\mathcal{K}$, we get the desired results. □

REMARK. In view of Lemma 10.1, it is seen that both of the quantities $\mathcal{K}(a)\,\mathcal{K}(b)$ and $\mathcal{L}(a)\,\mathcal{L}(b)$ are symmetric in a and b. It is also readily seen that the quantity $\mathcal{K}(a)^2$

is equal to the limit of $\varkappa(a)\varkappa(b)$ as $b \to a$, a fact which is also really an immediate consequence of the analyticity of $\varkappa(a)$ with respect to a.

Now, Lemma 10.1 just proved suggests that any product of $\varkappa(a)$'s with several values of a's be expressed as a linear form of the referring $\varkappa(a)$'s and their derivatives. For instance, if a, b, c are mutually distinct, then

$$\varkappa(a)\varkappa(b)\varkappa(c)$$

$$= \frac{\varkappa(a)}{(a-b)(a-c)} + \frac{\varkappa(b)}{(b-c)(b-a)} + \frac{\varkappa(c)}{(c-a)(c-b)}$$

while, if a and b are distinct, then

$$\varkappa(a)\varkappa(b)^2 = \frac{\varkappa(a) - \varkappa(b)}{(b-a)^2} + \frac{1}{b-a}\frac{d}{db}\varkappa(b).$$

In generalizing these examples, we shall deal with the product of any factors. Among them we begin with the product of simple factors and have the following theorem:

THEOREM 10.1. *If* a_ν $(\nu = 1, \ldots, n)$ *are mutually distinct, then*

$$\prod_{\nu=1}^{n} \varkappa(a_\nu) = (-1)^{n-1} \sum_{\nu=1}^{n} \frac{1}{\pi'(a_\nu; A_n)} \varkappa(a_\nu)$$

where $A_n = (a_1, \ldots, a_n)$ *and*

$$\pi(x;A_n)=\prod_{\nu=1}^{n}(x-a_\nu).$$

Proof. The case $n = 1$ is trivially true, since

$$\pi(x;A_1)=x-a_1,\quad \pi'(a_1;A_1)=1.$$

To proceed by induction, suppose that the case n is true. Then, by means of Lemma 10.1, we obtain, after multiplying $\chi(a_{n+1})$,

$$\prod_{\nu=1}^{n+1}\chi(a_\nu)$$

$$=(-1)^{n-1}\sum_{\nu=1}^{n}\frac{1}{\pi'(a_\nu;A_n)}$$

$$\cdot\frac{1}{a_{n+1}-a_\nu}(\chi(a_\nu)-\chi(a_{n+1})).$$

By the linear independence of $\{\chi(a_\nu)\}$, the coefficients of the linear expression in the right-hand member are uniquely determined. Accordingly, the coefficient of $\chi(a_\nu)$ with $1 \leqq \nu \leqq n$ is equal to

$$(-1)^{n-1}\frac{1}{\pi'(a_\nu;A_n)}\frac{-1}{a_\nu-a_{n+1}}=(-1)^{n}\frac{1}{\pi'(a_\nu;A_{n+1})},$$

as desired. On the other hand, by taking into account the

commutativity, we see that the last relation remains valid also for the coefficient of $\chi(a_\nu)$ with $\nu = n+1$. Thus, the proof by induction is complete. □

We remark here that, in connection with the fact mentioned at the end of the above proof, the relation

$$\sum_{\nu=1}^{n+1} \frac{1}{\pi'(a_\nu; A_{n+1})} = 0$$

is verified as a by-product, which is a classical elementary identity due to Euler.

In case some among a's coincide, the corresponding limit process may be applied. However, it will also be shown that an explicit expression can be deried. To see this, we first observe the power of a single factor. Then we obtain the following theorem which generalizes Lemma 10.1.

THEOREM 10.2. *For any integer* $h \geqq 0$,

$$\chi(a)^{h+1} = \frac{(-1)^h}{h!}\left(\frac{d}{da}\right)^h \chi(a).$$

Proof. Since the case $h = 0$ is trivial, we suppose $h > 0$. By the definition of $\mathcal{L}$-operator, we have

$$\mathcal{L}(a)^h f(z) = \frac{a^h}{(h-1)!}\int_I f(zt)\, t^{a-2}\left(\log\frac{1}{t}\right)^{h-1} dt,$$

whence follows after differentiating with respect to a the relation

$$\frac{d}{da}\mathcal{L}(a)^h f(z)$$

$$= \frac{a^h}{(h-1)!}\int_I f(zt)\,t^{a-2}\left(\frac{h}{a}-\log\frac{1}{t}\right)\left(\log\frac{1}{t}\right)^{h-1}dt$$

$$= \frac{h}{a}(\mathcal{L}(a)^h - \mathcal{L}(a)^{h+1})\ f(z).$$

Hence we get

$$\frac{d}{da}\mathcal{L}(a)^h = \frac{h}{a}(\mathcal{L}(a)^h - \mathcal{L}(a)^{h+1}),$$

which can be written into the form

$$\left(\frac{\mathcal{L}(a)}{a}\right)^{h+1} = -\frac{1}{h}\frac{d}{da}\left(\frac{\mathcal{L}(a)}{a}\right)^h.$$

This recurrence formula leads us inductively to the desired result. □

In the proof above, a recurrence formula for the $\mathcal{L}$-operator has been obtained:

$$\text{(10.2)} \qquad \mathcal{L}(a)^{h+1} = \mathcal{L}(a)^h - \frac{a}{h}\frac{d}{da}\mathcal{L}(a)^h.$$

In an alternative proof of Corollary 10.1 below, a recurrence formula will be crucial; cf. (10.3) below:

COROLLARY 10.1. *For any* $h \geqq 0$ *we have*

$$\mathcal{L}(a)^{h+1} = \sum_{i=0}^{h} (-1)^{i} \frac{a^{i}}{i!} \left(\frac{d}{da}\right)^{i} \mathcal{L}(a).$$

Proof. The desired result is just an expanded form of the relation given in Theorem 10.2 after putting $\mathcal{K}(a) = \mathcal{L}(a)/a$. In fact, the relation in Theorem 10.2 yields

$$\mathcal{L}(a)^{h+1}$$

$$= a^{h+1} \frac{(-1)^{h}}{h!} \left(\frac{d}{da}\right)^{h} \frac{\mathcal{L}(a)}{a}$$

$$= a^{h+1} \frac{(-1)^{h}}{h!} \sum_{i=0}^{h} \binom{h}{i} \left(\frac{d}{da}\right)^{h-i} \frac{1}{a} \left(\frac{d}{da}\right)^{i} \mathcal{L}(a)$$

$$= a^{h+1} \frac{(-1)^{h}}{h!}$$

$$\sum_{i=0}^{h} \frac{h!}{i!(h-i)!} (-1)^{h-1} \frac{(h-i)!}{a^{h-i+1}} \left(\frac{d}{da}\right)^{i} \mathcal{L}(a)$$

$$= \sum_{i=0}^{h} (-1)^{i} \frac{a^{i}}{i!} \left(\frac{d}{da}\right)^{i} \mathcal{L}(a).$$

Or, alternatively, the relation in Corollary 10.1 may be derived more directly as follows. By differentiating h times the defining equation of $\mathcal{L}(a)$ with respect to a and recalling the defining equation of $\mathcal{L}(a)^{\lambda}$, we get

$$\left(\frac{d}{da}\right)^{h} \mathcal{L}(a) f(z)$$

$$= \int_I f(zt)(at^{a-2}(\log t)^{h} + ht^{a-1}(\log t)^{h-1})\, dt$$

$$= \left((-1)^{h} a \frac{h!}{a^{h+1}} \mathcal{L}(a)^{h+1}\right.$$

$$\left. + h(-1)^{h-1} \frac{(h-1)!}{a^{h}} \mathcal{L}(a)^{h}\right) f(z)$$

$$= (-1)^{h} \frac{h!}{a^{h}} (\mathcal{L}(a)^{h+1} - \mathcal{L}(a)^{h}) f(z),$$

whence follows a recurrence formula of the type

$$(10.3) \quad \mathcal{L}(a)^{h+1} = \mathcal{L}(a)^{h} + (-1)^{h} \frac{a^{h}}{h!} \left(\frac{d}{da}\right)^{h} \mathcal{L}(a).$$

This leads to the desired result. □

Finally, we supplement the relations stated in Theorem

10.1 and Theorem 10.2, by considering the case where repeated factors appear in the product. In view of commutativity, such a product reduces to the standard form

$$\text{(10.4)} \qquad \Omega = \prod_{\nu=1}^{n} \chi(a_\nu)^{h_\nu+1}, \qquad h_\nu \geqq 0 \quad (\nu = 1, \ldots, n).$$

We now try to express the subproduct in Ω of the form

$$\chi(a)^{h+1}\chi(b)^{k+1} \qquad \text{with} \quad a \neq b$$

as a linear combination of $\chi(a)$, $\chi(b)$ and their derivatives.

THEOREM 10.3. *For* $a \neq b$ *and any pair of integers* $h, k \geqq 0$

$$\chi(a)^{h+1}\chi(b)^{k+1}$$

$$= \frac{(-1)^h}{k!}\sum_{i=0}^{h}\frac{(h+k-i)!}{i!(h-i)!}\,\frac{1}{(b-a)^{h+k-i+1}}\left(\frac{d}{da}\right)^{i}\chi(a)$$

$$+ \frac{(-1)^k}{h!}\sum_{j=0}^{k}\frac{(k+h-j)!}{j!(k-j)!}\,\frac{1}{(a-b)^{k+h-j+1}}\left(\frac{d}{db}\right)^{j}\chi(b).$$

Proof. By Theorem 10.2, we have

$$\Lambda(a)^{h+1}\chi(b)^{k+1} = \frac{(-1)^{h+k}}{h!k!}\left(\frac{d}{da}\right)^{h}\chi(a)\left(\frac{d}{db}\right)^{k}\chi(b).$$

To compute the right hand member, we use Lemma 10.1, the case $b \neq a$:

$$\mathcal{X}(a)\mathcal{X}(b) = \frac{1}{b-a}(\mathcal{X}(a) - \mathcal{X}(b)).$$

Applying $(\partial/\partial a)^h (\partial/\partial b)^k$ to the both members, we get

$$\left(\frac{d}{da}\right)^h \mathcal{X}(a) \left(\frac{d}{db}\right)^k \mathcal{X}(b)$$

$$= \sum_{i=0}^{h} \binom{h}{i} (-1)^{h-i} \left(\left(\frac{\partial}{\partial b}\right)^{h+k-i} \frac{1}{b-a}\right) \left(\frac{d}{da}\right)^i \mathcal{X}(a)$$

$$+ \sum_{j=0}^{k} \binom{k}{j} (-1)^{k-j} \left(\left(\frac{\partial}{\partial a}\right)^{k+h-j} \frac{1}{a-b}\right) \left(\frac{d}{db}\right)^j \mathcal{X}(b).$$

Then, the desired result is obtained by using obvious equality

$$\left(\frac{\partial}{\partial v}\right)^m \frac{1}{v-u} = \frac{(-1)^m m!}{(v-u)^{m+1}}. \qquad \square$$

COROLLARY 10.2. *For $a \neq b$ and any pair of integers $h, k \geq 0$,*

$$\left(\frac{d}{da}\right)^h \mathcal{X}(a) \left(\frac{d}{db}\right)^k \mathcal{X}(b)$$

$$= (-1)^k h! \sum_{i=0}^{h} \frac{(h+k-i)!}{i!(h-i)!} \frac{1}{(b-a)^{h+k-i+1}} \left(\frac{d}{da}\right)^i \chi(a)$$

$$+ (-1)^h k! \sum_{j=0}^{k} \frac{(k+h-j)!}{j!(k-j)!} \frac{1}{(a-b)^{k+h-j+1}} \left(\frac{d}{db}\right)^j \chi(b).$$

Proof. Use Theorem 10.2. □

Using Theorem 10.3, its Corollary 10.2 and Theorem 10.2 repeatedly, we see that the product

$$\Omega = \prod_{\nu=1}^{n} \chi(a_\nu)^{h_\nu + 1}$$

is explicitly expressed as a linear form such that the arguments are $\chi(a_\nu)$ and their derivatives of order less than degrees $h_\nu + 1$ for $\nu = 1, \ldots, n$. Consequently, for the operators $\kappa(a_1), \ldots, \kappa(a_n)$ and an arbitrary polynomial

$$P(x_1, \ldots, x_n) = \sum_{\alpha_1, \ldots, \alpha_n} c_{\alpha_1 \ldots \alpha_n} x_1^{\alpha_1} \ldots x_n^{\alpha_n},$$

we have an algorithm of expressing the operator

$$\Omega = P(\mathcal{K}(a_1), \ldots, \mathcal{K}(a_n))$$

as a linear combination of $\{\mathcal{K}(a_\nu)\}_{\nu=1}^{n}$ and their derivatives, where the highest order of the derivatives of each $\mathcal{K}(a_\nu)$ is less than the degree of $P(x_1, \ldots, x_n)$ regarded as a polynomial of x_ν.

§ 11. Analytic prolongation

The main purpose of the present section is to deal with analytic prolongation of the operator $\mathcal{L}(a)^\lambda$ and to indicate its relation to a differential operator; cf. Komatu [20].

The operator $\mathcal{L}(a)^\lambda$ was initially defined for $\lambda \geqq 0$ and $a > 0$. The main reason of this restriction consists in preserving its integral representation.

However, the Taylor expansion of $\mathcal{L}^\lambda(a)$ for $f \in \mathcal{F}$ converges surely in E for any pair of complex values of λ and a, provided every member of $\{\alpha_\nu(a)^\lambda\}_{\nu=2}^{\infty}$ has definite finite value. In fact, let the Taylor expansion of $f \in \mathcal{F}$ be given by

$$f(z) = \sum_{\nu=1}^{\infty} c_\nu z^\nu \quad \text{with} \quad c_1 = 1.$$

Substitution followed by termwise integration then yields

$$\mathcal{L}(a)^{\lambda} f(z) = \sum_{\nu=1}^{\infty} \alpha_{\nu}(a)^{\lambda} c_{\nu} z^{\nu}$$

where

$$\{\alpha_{\nu}(a)\}_{\nu=1}^{\infty} = \left\{\frac{a}{a+\nu-1}\right\}_{\nu=1}^{\infty}$$

is the moment sequence with respect to the probability measure $\sigma(t\,;\,a)$. Thus, $\mathcal{L}(a)^{\lambda}$ will be analytically prolongable into such range of λ and a, if it is submitted to the inevitable the condition that $d\sigma_{\lambda}(t\,;\,a) \geqq 0$ cannot be preserved.

We begin with the following theorem:

THEOREM 11.1. *The operator* $\mathcal{L}(a)^{\lambda}$ *on* $\mathcal{F}$ *is analytically prolongable with respect to a pair of parameters* λ *and* a *within single - valuedness into the whole complex pair cut along the negative real axis on the* a *- plane.*

Proof. The analytic prolongability follows readily from the series expansion of $\mathcal{L}(a)^{\lambda} f$. On the negative real axis of the a-plane $\mathcal{L}(a)^{\lambda}$ behaves analytically, provided $a \neq -1, -2, \ldots$ if $\operatorname{Re}\lambda > 0$, $a \neq 0$ if $\operatorname{Re}\lambda < 0$, and $a \neq 0, -1, -2, \ldots$ if $\operatorname{Re}\lambda = 0$. However, the prolonged operator then shows multi-valuedness when a crosses the negative real axis, unless λ coincides with an integer. □

Now, the original integral representation of $\mathcal{L}(a)^{\lambda}$

shows that the normalization $f'(0) = 1$ is inessential and further it is applicable to any holomorphic function of a wider class without any normalization at the origin, provided Re $a > 1$. The series expansion then becomes

$$\mathcal{L}(a)^{\lambda} f(z) = \sum_{\nu=0}^{\infty} \left(\frac{a}{a + \nu - 1} \right)^{\lambda} c_{\nu} z^{\nu}$$

for

$$f(z) = \sum_{\nu=0}^{\infty} c_{\nu} z^{\nu}.$$

The operator $\mathcal{L}(a)^{\lambda}$ on this wider class will have a further singularity $a = 1$ if $\lambda > 1$.

In the following lines, we shall mainly observe $\mathcal{L}(a)^{\lambda}$ in the whole range of complex pairs of λ and a except real negative values of a. Any equality relation of analytic character with respect to these parameters will then remain valid after prolongation. Here, in connection with the fact mentioned above, we state here a relation between the quantities $\mathcal{L}(a + 1)^{\lambda}(f(z)/z)$ and $\mathcal{L}(a)^{\lambda} f(z)$:

THEOREM 11.2. *For any* $f \in \mathcal{F}$

$$\mathcal{L}(a + 1)^{\lambda} \frac{f(z)}{z} = \left(\frac{a + 1}{a} \right)^{\lambda} \frac{1}{z} \mathcal{L}(a)^{\lambda} f(z). \tag{11.1}$$

Proof. In view of analyticity, we may suppose $a > 0$. Direct calculation yields

$$\mathcal{L}(a+1)^{\lambda}\,\frac{f(z)}{z}$$

$$= \frac{(a+1)^{\lambda}}{\Gamma(\lambda)} \int_I t^{a-1}\left(\log\frac{1}{t}\right)^{\lambda-1} \frac{f(zt)}{zt}\,dt$$

$$= \left(\frac{a+1}{a}\right)^{\lambda} \frac{1}{z}\,\frac{a^{\lambda}}{\Gamma(\lambda)} \int_I t^{a-2}\left(\log\frac{1}{t}\right)^{\lambda-1} f(zt)\,dt$$

$$= \left(\frac{a+1}{a}\right)^{\lambda} \frac{1}{z}\,\mathcal{L}(a)^{\lambda} f(z).$$

The desired relation can be alternatively verified by means of series expansion. □

Finally, we introduce a differential operator $\Theta(a)$ depending on a parameter a by

(11.2) $$\Theta(a) = \frac{1}{a}(\theta + a - 1), \qquad \theta = \frac{d}{d\log z}.$$

Then, it is readily seen that $f \in \mathcal{F}$ implies $\theta f \in \mathcal{F}$ and hence $\Theta(a) f \in \mathcal{F}$. The definition of Θ may also be represented as in the manner

$$\Theta(a) f(z) = \frac{1}{a z^{a-1}}\,\theta(z^{a-1} f(z)) \qquad (f \in \mathcal{F}).$$

The particular operator $\mathcal{L}(1)$ is the integration with respect to $\log z$, that is, it is the inverse operator of $\theta = \Theta(1)$. In fact, θ commutes with any $\mathcal{L}$ and for any $f \in \mathcal{F}$ we have

$$\mathcal{L}(1)\ \theta\ f(z) = \int_I zf'(zt)\,dt = f(z).$$

This relation is generalized as follows:

THEOREM 11.3. *The operator $\mathcal{L}(a)$ is the inverse to $\Theta(a)$.*

Proof. In view of analyticity, we may suppose $a > 0$. Since θ is commutable with any $\mathcal{L}$, so is $\Theta(a)$ also. For any $f \in \mathcal{F}$ we have

$$\mathcal{L}(a)\ \Theta(a) f(z)$$

$$= \int_I (zf'(zt)\,t^{a-1} + (a-1) f(zt)\,t^{a-2})\,dt$$

$$= \int_I \frac{\partial}{\partial t}(f(zt)\,t^{a-1})\,dt = f(z).$$

Or, the relation can be alternatively be verified by making use of series form. □

It has been shown in Theorem 7.2 that the relation

$$\theta\ \mathcal{L}(a)^{\lambda+1} = a\,\mathcal{L}(a)^{\lambda} - (a-1)\mathcal{L}(a)^{\lambda+1}$$

holds for any $a > 0$ and $\lambda \geqq 0$. This relation may be used in verifying the analytic prolongeability with respect to λ. It is noted, by the way, that the relation is an immediate consequence of Theorem 11.3. In fact, $\Theta(a)\ \mathcal{L}(a) = \text{id}$ implies the relation $\Theta(a)\ \mathcal{L}(a)^{\lambda+1} = \mathcal{L}(a)^{\lambda}$ which is readily brought into the desired form.

On the other hand, by referring to the operator $\Theta(a)$, any operator $\mathcal{L}(a)^{\lambda}$ with $\operatorname{Re}\lambda < 0$ can be expressed in terms of $\mathcal{L}(a)^{\mu}$ with $\operatorname{Re}\mu > 0$, a fact which may be stated as in the following theorem:

THEOREM 11.4. *The operator* $\mathcal{L}(a)^{\lambda}$ *with* $\operatorname{Re}\lambda < 0$ *is expressible in terms of* $\mathcal{L}(a)^{\mu}$ *with* $\operatorname{Re}\mu > 0$ *in the form*

$$\mathcal{L}(a)^{\lambda} = \Theta(a)^{m}\,\mathcal{L}(a)^{\lambda+m}$$

where m *is any positive integer satisfying* $m > [-\operatorname{Re}\lambda]$.

Proof. By making use of the commutativity between $\Theta(a)$ and $\mathcal{L}(a)^{\lambda}$, the desired relation follows readily from Theorem 11.3. □

COROLLARY 11.1. *If, in particular,* $\operatorname{Re} a > 0$, *the relation stated in the Theorem* 11.4 *can be brought into the form*

$$\mathcal{L}(a)^{\lambda} f(z)$$

$$= \frac{a^{\lambda}}{\Gamma(\lambda+m)} \sum_{k=0}^{m} \binom{m}{k} (a-1)^{m-k}$$

$$\cdot \int_I \left(z\frac{d}{dz}\right)^{k} f(zt)\, t^{a-2} \left(\log\frac{1}{t}\right)^{\lambda+m-1} dt \quad (f \in \mathcal{F}).$$

Proof. The development can be readily obtained by writing the expression of $\mathcal{L}(a)^{\lambda+m} f(z)$ and then expanding the expression $\Theta(a)^{m}$ in terms of θ. □

We note that Theorem 11.3 leads us naturally to define $\Theta(a)^{\lambda}$ with any complex order λ. In fact, the interrelation

$$\Theta(a)\,\mathcal{L}(a) = \mathrm{id}$$

suggests that it is appropriate to define $\Theta(a)^{\lambda}$ by means of

$$\Theta(a)^{\lambda} = \mathcal{L}(a)^{-\lambda}.$$

In particular, $\theta^{\lambda} = \mathcal{L}(1)^{-\lambda}$ corresponds to the differentiation of complex order with respect to log z.

Finally, we supplement a short remark. If $a > 0$, the sequence of moments $\{\alpha_{\nu}(a)^{-\lambda}\}_{\nu=1}^{\infty}$ is increasing for $\lambda > 0$. Hence, the operator $z^{-1}\Theta(a)^{-\lambda}$ with $a > 0$ and $\lambda > 0$ is regarded as a special case of the Gel'fond-Leont'ev derivative.

§ 12. Functional equations

Most of the relations concerning the operator $\mathcal{L}$ of the form

$$\mathcal{L} f(z) = F(z), \qquad F \in \mathcal{F},$$

may be regarded as functional equations with respect to the unknown function $f(z) \in \mathcal{F}$.

A typical problem is to solve the equation

$$\mathcal{L}(a)^{\lambda} f(z) = F(z) \tag{12.1}$$

with respect to $f \in \mathcal{F}$, where $F \in \mathcal{F}$ is a known function.

Since we have shown in Theorem 11.3 that $\mathcal{L}(a)$ is inverse to $\Theta(a)$, the equation (12.1) is solvable explicitly provided $\Theta(a)^{-\lambda}$ is given in explicit form.

We begin with the simplest case, where $a = 1$ in the equation (12.1).

THEOREM 12.1. *The functional equation*

$$\mathcal{L}(1)^m f(z) = F(z) \tag{12.2}$$

with unknown function $f(z)$ *satisfying* $f \in \mathcal{J}$, *where* m *is a given positive integer and* F *is a known function satisfying* $F \in \mathcal{J}$, *is solved explicitly by*

$$f(z) = \left(\frac{d}{d\log z}\right)^m F(z). \tag{12.3}$$

Proof. First, in view of Theorem 11.3, the solution of (12.2) is given by

$$f(z) = \Theta(1)^m F(z).$$

On the other hand, since we see that

$$\Theta(1) = \frac{d}{d\log z},$$

the desired result (12.3) follows. □

Next, we shall state a similar result in slightly generalized form:

THEOREM 12.2. *The functional equation*

$$(12.4) \qquad \mathcal{L}(a)^m f(z) = F(z)$$

with unknown function $f(z)$ *satisfying* $f \in \mathcal{J}$ *where* m *is a given positive integer and* $F(z)$ *is a known function satisfying* $F \in \mathcal{J}$, *is solved explicitly by*

$$(12.5) \qquad f(z) = \left(\frac{1}{a}\left(\frac{d}{d\log z} + a - 1\right)\right)^m F(z)$$

or

$$(12.6) \qquad f(z) = \frac{1}{a^m}\sum_{k=0}^{m}\binom{m}{k}(a-1)^{m-k}\left(z\frac{d}{dz}\right)^k F(z).$$

Proof. The proof proceeds quite similarly as the previous Theorem 12.1. The solution of the equation (12.4) can be obtained in the form

$$f(z) = \theta(a)^m F(z)$$

which is exactly (12.5). In expanding the right-hand member of the expression

$$\theta(a)^m = \frac{1}{a^m}(\sigma + a - 1)^m$$

by means of binomial series in θ, we obtain the desired result (12.6). □

Finally, by taking into account of the analyticity of the expressions, the similar method will apply also to the case where the restriction that m is an integer is removed. Namely, we have the following Theorem:

THEOREM 12.3. *The functional equation*

$$\mathcal{L}(a)^{\lambda} f(z) = F(z) \tag{12.7}$$

with unknown function $f(z)$ *satisfying* $f \in \mathcal{F}$ *where* $a > 1$ *is a fixed number,* λ *is a given positive real number and* $F(z)$ *is a known function satisfying* $F \in \mathcal{F}$, *is solved explicitly by*

$$f(z) = \Theta(a)^{\lambda} F(z) \tag{12.8}$$

where $\Theta(a)^{\lambda}$ *may be expanded in series of* $\theta = d/d\log z$ *in the following form*

$$\Theta(a)^{\lambda} = \left(\frac{a-1}{a}\right)^{\lambda} \sum_{\nu=0}^{\infty} \binom{\lambda}{\nu} \frac{1}{(a-1)^{\nu}} \theta^{\nu}. \tag{12.9}$$

Proof. The proof is similar to that of Theorem 12.2. □

Now, for the sake of brevity, we restrict ourselves to the case where $\lambda = m$ is a positive integer. In Theorem 11.2 we have derived the relation

$$\mathcal{L}(a+1)^{m} \frac{f(z)}{z} = \left(\frac{a+1}{a}\right)^{m} \frac{1}{z} \mathcal{L}(a)^{m} f(z)$$

where $f \in \mathcal{F}$ and we suppose here $a > 0$.

With reference to this relation, we state here the following theorem:

THEOREM 12.4. *The functional equation*

$$\mathcal{L}(a)^{m}\frac{f(z)}{z}=\left(\frac{a}{a-1}\right)^{m}\frac{1}{z}\mathcal{L}(a-1)^{m}F(z)$$

with unknown function $f(z)$ *satisfying* $f \in \mathcal{F}$ *where* m *is a positive integer,* a *satisfies* $a > 2$ *and* $F(z)$ *is a known function satisfying* $F \in \mathcal{F}$, *may be explicitly solved in the form*

$$f(z)$$

$$=\frac{1}{(m-1)!}\sum_{k=0}^{m}\binom{m}{k}(a-1)^{m-k}$$

$$\cdot\int_{I} t^{a-3}\left(\log\frac{1}{t}\right)^{m-1}\left(\frac{d}{d\log z}\right)^{k}\left(\frac{F(zt)}{zt}\right)dt\,.$$

Proof. The proof proceeds quite similarly as in the previous Theorems. In this Theorem the use is made, according to the circumstances, $a-1$ instead of a and m instead of λ in the Theorem 11.2. □

It is noticed that the solution may be also written in the form

$$\frac{f(z)}{z} = \sum_{k=0}^{m} \binom{m}{k} (a-1)^{-k} \left(\frac{d}{d \log z} \right)^{k} \left(\frac{1}{z} \mathcal{L} (a-1)^{m} F(z) \right).$$

PART II. Distortion Theorems

In the present part, we shall deal with distortions of various kinds on miscellaneous functionals. The leading tools of attack are the linear integral operators introduced and investigated in detail in the former part.

Through the actions of the operators which are further generalized or specialized, properties of various kinds of functionals will be clarified.

Chapter 4. Distortions on univalent functions

§ 13. Maximum modulus

In the former half of this monograph, we have dealt with different faces of a class of linear integral operators; especially, their own character of addiivity, namely, character of semiring structure, several kinds of relations to fractional calculus and integration operator, the properties of their products as well as their analytic prolongeability.

These operators will play basic roles in the subsequent chapters. In fact, several kinds of functionals, we deal with estimations by means of various distortion inequalities for these funtionals as integral operators acting on a class of functions. A typical problem states:

Given a functional $\Omega[f]$ defined on the class of functions $\mathcal{F} = \{f\}$ which consists of analytic functions f holomorphic in the unit disk E and normalized by $f(0) = f'(0) - 1 = 0$, we consider an integral operator $\mathcal{L}$ defined on $\mathcal{F}$. Then, the problem is to obtain estimations of $\Omega[\mathcal{L} f]$ for for $\{\mathcal{L} f\}_{f \in \mathcal{F}}$ in terms of known quantities.

For each pair of a functional $\Omega[f]$ and an operator $\mathcal{L}$, distortion inequalities will be derived.

In the present chapter, we consider the maximum modulus of $F \in \mathcal{F}$ as a basic model of such functionals. Let the maximum modulus of a function F along a concentric circumference $\{|z| = r\}$ be denoted by

$$(13.1) \qquad M_r[F] := \max_{|z|=r} |F(z)| \qquad (F \in \mathcal{F}),$$

where $r \in (0, 1)$ is a parameter arbitrarily fixed; cf. Komatu [5, 14].

We consider a functional $\mathcal{L}: \mathcal{F} \to \mathcal{F}$ defined by

$$(13.2) \qquad \mathcal{L}f(z) := \int_I \frac{f(zt)}{t}\, d\sigma(t),$$

where σ is a probability measure supported by the unit interval $I = [0, 1]$. In the following lines the Dirac measure concentrated at the single point 1 is occasionally exceptional, and we denote it by σ^*. Every $f \in \mathcal{F}$ is transformed by $\mathcal{L}$ with $\sigma = \sigma^*$ into itself, while the particular function z is always transformed into itself by $\mathcal{L}$ with any σ.

THEOREM 13.1. *For any $f \in \mathcal{F}$ the monotonicity*

$$(13.3) \qquad M_r[\mathcal{L}f] \leqq M_r[f]$$

holds. Unless $\sigma = \sigma^*$, the equality sign in (13.3) *is realized for a certain $r \in (0, 1)$ if and only if $f(z) \equiv z$.* .

Proof. We may assume $\sigma \neq \sigma^*$. The maximum principle yields

$$\left|\frac{f(zt)}{zt}\right| \leq \max_{|z|=r}\left|\frac{f(z)}{z}\right| = \frac{M_r[f]}{r} \qquad (t \in [0, 1])$$

for any z with $|z| \leq r < 1$. Hence, we get

$$\left|\frac{\mathcal{L}f(z)}{z}\right| \leq \int_I \left|\frac{f(zt)}{zt}\right| d\sigma(t) \leq \frac{M_r[f]}{r}$$

and consequently

$$M_r[\mathcal{L}f] \leq M_r[f].$$

The equality holds if and only if $|f(zt)/zt| = M_r[f]/r$ at every t with $d\sigma(t) > 0$ and hence $f(z) = z$, unless $\sigma = \sigma^*$. □

THEOREM 13.2. *For any f, $g \in \mathcal{F}$, we have*

$$M_r[\mathcal{L}f - \mathcal{L}g] \leq M_r[f - g] \int_I t\, d\sigma(t).$$

Proof. Since f and g satisfy the same normalization at the origin, the quantity $(f(zt) - g(zt))/t^2$ may be regarded as a function of a complex variable t holomorphic throughout $\{|t| \leq 1\}$. The maximum principle applied to this quantity qua function of t yields

$$\max_{|t|\leq 1}\left|\frac{f(zt) - g(zt)}{t^2}\right| = \max_{|z|=r}|f(z) - g(z)|,$$

whence follows for $t \in I$

$$|f(zt) - g(zt)| \leqq t^2 M_r[f - g].$$

Consequently, we obtain

$$\begin{aligned} &|\mathcal{L}f(z) - \mathcal{L}g(z)| \\ &\leqq \int_I \frac{|f(zt) - g(zt)|}{t} d\sigma(t) \\ &\leqq M_r[f - g] \int_I t\, d\sigma(t), \end{aligned}$$

which yields the desired result. □

The Theorem 13.2 just proved can be slightly generalized. Namely, by refining a classification in $f \in \mathcal{F}$ via normalization at the origin, the difference of two $\mathcal{L}f$ in the same class can be estimated more precisely. For instance, we state here the following theorem:

THEOREM 13.3. *If $f, g \in \mathcal{F}$ possess the Taylor coefficients in common up to the power of k, then*

$$M_r[\mathcal{L}f - \mathcal{L}g] \leqq M_r[f - g] \int_I t^k d\sigma(t). \tag{13.4}$$

Proof. The proof proceeds quite similarly to that of the previous Theorem 13.2. For any fixed z with $|z| = r < 1$, the quantity $(f(zt) - g(zt))/t^{k+1}$ may be regarded as function of a complex variable t holomorphic on $\{|t| \leqq 1\}$. The maximum principle applied to this function yields

$$\max_{|t|\leq 1}\left|\frac{f(zt)-g(zt)}{t^{k+1}}\right| = \max_{|z|=1}|f(z)-g(z)|,$$

whence follows

$$|f(zt)-g(zt)| \leq t^{k+1} M_r[f-g]$$

$$|\mathcal{L}f(z)-\mathcal{L}g(z)| \leq \int_I \frac{|f(zt)-g(zt)|}{t}\,d\sigma(t)$$

$$\leq M_r[f-g]\int_I t^k\,d\sigma(t),$$

which is the desired result. □

With respect to the quotiemt of $f, g \in \mathcal{J}$, instead of their difference, we see that $(f/g)(0) = 1$ and

$$\left| zt\frac{f(zt)}{g(zt)}\right| \leq \max_{|z|=r}\left| z\frac{f(z)}{g(z)}\right|$$

$$= M_r\left[z\frac{f}{g}\right] \quad (t \in [0, 1])$$

for any z with $|z| = r < 1$. Hence, we obtain

$$M_r\left[\mathcal{L}\left(z\frac{f}{g}\right)\right] \leq M_r\left[z\frac{f}{g}\right],$$

which is no more than the estimation derived in Theorem 13.1 applied to zf/g instead of f.

Differently from the difference, for the estimation on the quotient one can obtain quite little.

Next, in relation with (13.1), we introduce for a pair $f, g \in \mathcal{F}$ two quantities defined by

$$M(r;\ a,\ \lambda,\ \mu) = \max_{|z|=r} |\mathcal{L}(a)^{\lambda} f(z) - \mathcal{L}(a)^{\mu} g(z)|,$$

$$N(r;\ a,\ \lambda) = \max_{|z|=r} |\mathcal{L}(a)^{\lambda} f(z) - z|.$$

Then, we obtain the following theorem:

THEOREM 13.4. *For any* $f, g \in \mathcal{F}$ *the quantity* $M(r;\ a,\ \lambda + \delta,\ \mu + \delta)$ *decreases with respect to* δ. *More precisely, for* $\delta' > \delta \geqq 0$ *we have*

$$\left(\frac{a+1}{a}\right)^{\delta'} M(r;\ a,\ \lambda + \delta',\ \mu + \delta')$$

$$\leqq \left(\frac{a+1}{a}\right)^{\delta} M(r;\ a,\ \lambda + \delta,\ \mu + \delta).$$

Proof. Since both $\mathcal{L}(a)^{\lambda} f$ and $\mathcal{L}(a)^{\mu} g$ belong to the class $\mathcal{F}$, we have in view of Theorem 13.2

$$M(r;\ a,\ \lambda + \delta,\ \mu + \delta)$$

$$\leqq M(r;\ a,\ \lambda,\ \mu) \int_I t\, d\sigma_{\delta}(t;\ a).$$

The last factor of the right-hand member of this equality is equal to

$$\int_I t\, d\sigma_{\delta}(t;\ a) = \frac{a^{\delta}}{\Gamma(\delta)} \int_I t^{a} \left(\log \frac{1}{t}\right)^{\delta - 1} dt$$

$$= \left(\frac{a}{a+1}\right)^{\delta}.$$

Let $0 \leqq \delta < \delta'$. Then, by replacing λ, μ and δ in the above inequality by $\lambda + \delta$, $\mu + \delta$ and $\delta' - \delta$, respecttively, we obtain the desired result. □

COROLLARY 13.1. *For any $f \in \mathcal{S}$ the quantity*

$$\left(\frac{a+1}{a}\right)^{\delta} N(r;\, a,\, \lambda + \delta)$$

decreases with respect to $\delta \geqq 0$.

Proof. Since $\mathcal{L}(a)^{\mu} z$ becomes z for any μ, the quantity $M(r;\, a,\, \lambda,\, \mu)$ reduces to $N(r;\, a,\, \lambda)$ provided $g(z) = z$. Hence, the assertion follows from Theorem 13. 4 by only substituting $g(z) = z$. □

By the way, from the Corollary 13.1 just shown, that

$$\left(\frac{a+1}{a}\right)^{\delta} N(r;\, a,\, \lambda + \delta) \leqq N(r;\, a,\, \lambda).$$

If we replace here both λ and δ by $\lambda/2$, we get

$$N(r;\, a,\, \lambda) \leqq \left(\frac{a}{a+1}\right) N\left(r;\, a, \frac{\lambda}{2}\right).$$

In view of this inequality, we see that the first limit relation mentioned in Theorem 5.3 is again verified.

Next, we shall refer to a theorem of Rogosinski [2]. He obtained a precision of a theorem of Schwarz [1] for a class of functions φ in $\mathcal{J}$ satisfying a condition that besides $\varphi(0) = 0$ every φ attains real values alone on the real diameter. Here, we shall mention its application. Though the result will not be in direct connection with the estimation of maximum modulus but since the problem lies in the category of Schwarz theorem basic in the theory of bounded functions; For the general theory on bounded functions, cf. Schur [1].

We mention the theorem of Rogosinski as a Lemma:

LEMMA 13.1. *Let* φ *be a function holomorphic in the unit disk* E *which satisfies* $\varphi(0) = 0$, $|\varphi(z)| < 1$ $(z \in E)$ *and attains real values alone along the diameter lying on the real axis. Then, for each value* $z \in E$ *the image point* $\varphi(z)$ *belongs to the closed lune enclosed by two minor circular arcs through* $1, z, -z$ *and* $-1, -z, z$; *in particular if* z *is real, then the lune is understood to be the segment between* z *and* $-z$. *If* z *is not real, the boundary point of this lune is attained only by the functions of the form*

$$(13.5) \qquad \pm \varphi_0(z; \tau) = \pm z \frac{z - \tau}{\tau z - 1}, \qquad -1 \leqq \tau \leqq 1.$$

Proof. First, suppose that $\varphi(\tau) \geqq 0$ for real $\tau \in (0 < |\tau| < 1)$. Then, the function defined by

$$\Psi(z) = \frac{\varphi(\psi(z; \tau)/z)}{\psi(z; \tau)/z}$$

where

$$\psi(z\,;\,\tau) = \frac{z - \tau}{\tau z - 1}$$

is holomorphic in E and satisfies

$$|\Psi(z)| \leq 1, \qquad \Psi(0) = \frac{\varphi(\tau)}{\tau} \leq 1.$$

Hence, in view of the principle of Lindelöf [1], the point $\Psi(z)$ does belong to a closed disk $K_{|\zeta|}(f(\tau)/\tau))$ which is the image of $\{|z| \leq |\zeta|\}$. Since the inverse function ψ^{-1} of ψ with respect to z coincides with ψ itself, we have

$$\Psi(\psi^{-1}(z\,;\,\tau)) = \Psi(\psi(z\,;\,\tau))$$

$$\in K_{|\psi(z\,;\,\tau)|}\left(\frac{\varphi(\tau)}{\tau}\right).$$

Since $\Psi(0) \equiv \varphi(\tau)/\tau$ is contained in I, the point $\varphi(z)/z$ belongs to the union of $K_{|\psi(z\,;\,\tau)|}(t)$ over $t \in I$. Now, together with $\varphi(z)$, the function $-\varphi(z)$ also satisfies the assumption, and moreover $\pm\varphi(-z)$ satisfies the assumption with respect to $-\tau$ instead of τ. Furthermore, among four functions $\pm\varphi(z)$ and $\pm\varphi(-z)$, one satisfies the assumption with τ or $-\tau$, and $\psi(-z\,;\,-\tau) \equiv -\psi(z\,;\,\tau)$. Hence, $\pm\varphi(z)/z$ and $\pm\varphi(-z)/(-z)$ for every function f also belong to

$$\{K_{|\psi(z\,;\,\tau)|} \cup K_{-|\psi(z\,;\,\tau)|}\}.$$

Accordingly, $\varphi(z)$ belongs to the closed lune

$$\{z\, K_{|\psi(z\,;\,\tau)|}\}.$$

Now, if τ runs over $(-\infty, \infty]$, two points $\pm\,\psi(z\,;\,\tau)$ describe the circumferences through ± 1, ∓ 1, $\mp z$, respectively, whose parts corresponding to $[-1, +1]$ are the minor arcs with ± 1 and ∓ 1 as end-points and involving $\mp z$. Since the centers of these circumferences lie on the imaginary axis, their points nearest to the origin are the intersection points with the imaginary axis, whose distance from the origin is equal to $|\psi(z\,;\,\tau)|$. However, these circumferences pass through the points $\mp 1/z$ corresponding to $\tau = \infty$. Hence, we arrive at the conclusion of the Lemma.

With respect to the extremal function, if

$$\frac{\varphi_0(z_0)}{z_0} \in \partial K_{|\psi(z_0\,;\,\tau)|}$$

then it is shown that φ_0 must be of the form

$$\varphi(z) = z\,\frac{\varepsilon\psi(z\,;\,\tau) - t}{\varepsilon\, t\, \psi(z\,;\,\tau) - 1}\,;$$

$$0 < |t| \leqq 1, \quad \varepsilon\psi(z_0\,;\,\tau) = \pm\, i\, |\psi(z_0\,;\,\tau)|.$$

Since $\varphi_0(z)$ remains real for every real z, ε must be real, that is, $\varepsilon = \pm 1$, and hence

$$\varphi_0(z) = \mp z\left(z - \frac{\pm\tau - t}{\pm 1 - t\tau}\right)\Big/\left(\frac{\pm\tau - t}{\pm 1 - t\tau} z - 1\right).$$

Here $(\pm\tau - t)/(\pm 1 - t\tau)$ is a real number whose absolute value does not exceed the unity, so that by denoting it τ again, we see that the function (13.5) shows the extremality with a suitable τ. □

Lemma 13.1 can be diversely applicable. As its simple application, we shall here mention the following theorem:

THEOREM 13.5. *Let f be holomorphic in E and satisfy $f(0) = 0$, $|f(z)| < 1$ in E. Further, let f attain real values alone along the real diameter. Then for every point $z \in E$, the point of the range of $\mathcal{L}^{\lambda} f(z)$ generated by a real probability measure belongs to the lune bounded by two minor arcs with endpoints $\pm z$ passing through $1, z, -z$ and $-1, -z, z$; the lune being understood to be the segment between $\pm z$, if z is real. In particular, it satisfies*

$$|\mathrm{Re}\, \mathcal{L}^{\lambda} f(z)| \leqq |\mathrm{Re}\, z|$$

and

$$|\mathrm{Im}\, \mathcal{L}^{\lambda} f(z)| \leq |\mathrm{Im}\, z^{*}|,$$

where z^{*} denotes the highest point of the lunes.

Proof. If $\mathcal{L}^{\lambda} f(z)$ is generated by a real probability measure, then it satisfies the same condition as $f(z)$. Namely,

$$\mathcal{L}^{\lambda} f(0) = 0$$

and in view of Theorem 13.1

$$|\mathcal{L}^{\lambda} f(z)| \leq M_r[f(z)] < 1.$$

Further, as readily seen from its integral representation, it attains real value alone along the real diameter. Hence the conclusion follows from Lemma 13.1. The last part is evident since the range is contained in the rectangle

$$\{|\mathrm{Re}\ \zeta| \leq |\mathrm{Re}\ z|,\ |\mathrm{Im}\ \zeta| \leq |\mathrm{Im}\ z^{*}|\}. \qquad \square$$

In general, the problem on maximum modulus of analytic functions are very popular in the function theory. Accordingly, the attack has been made in various ways. It is especially indispensable as an effective method. It may be fairly said that it plays a central role in this field.

In § 8 we have introduced a class of functions for which a detailed discussion has been made. Several distortion theorems can be transferred to this case. For instance, we state here a simple theorem:

THEOREM 13.6. *Let* F_1 *and* F_2 *of the same form as* F *in Lemma* 8.2 *possess the coefficients corresponding to* B_κ *with* $1 \leq \kappa \leq k$ *in common. Then we have*

$$M_r[\mathcal{L}F_1 - \mathcal{L}F_2] \leq M_r[F_1 - F_2] \int_I t^k \, d\sigma(t).$$

Proof. The assertion has been essentially established in Theorem 8.3. □

§ 14. Classes related to uivalent functions

In the present section, we shall be mainly concerned in the classes related to univalent functions.

The theory of univalent functions has been variously developed. It plays a fundamental role in the theory of conformal mappings. Though we shall deal here with simple problems only, the whole aspect is quite wide and deep.

In the first place, we begin with some subclasses relating to the subject.

Let $\mathcal{J}^+$ be a subclass of $\mathcal{J}$ which consists of functions $f \in \mathcal{J}$ satisfying

$$\text{Re}\,\frac{f(z)}{z} > 0 \quad \text{in } E \tag{14.1}$$

and let $\mathcal{Y}$ denote its proper subclass which consists of $f \in \mathcal{J}^+$ satisfying

$$\text{Re}\,\frac{f(z)}{z} > \frac{1}{2}. \tag{14.2}$$

cf. Komatu [5]. In general, the so-called Carathéodory class

$$P(\alpha) = \{p\}$$

which consists of holomorphic functions p satisfying

$$\operatorname{Re} p(z) > \alpha$$

and normalized by $p(0) = p'(0) - 1 = 0$ at the origin will appear henceforth very often; for instance, §§ 27 and 28.

In particular, as shown by Strohhäcker [1], the class $\mathscr{K}$ consisting of convex mappings f belongs to $\mathscr{H}$: $\mathscr{K} \subset \mathscr{H}$. For general theory of univalent functions, cf. Goodman [1], Pommerenke [1].

We first consider here the class $\mathscr{H}$. It is noticed, in general, that between any two Carathéodory classes there exists a relation, i. e., if $f \in P(\alpha)$, $g \in P(\beta)$, then

$$\frac{f - \alpha z}{1 - \alpha} \Leftrightarrow \frac{g - \beta z}{1 - \beta}.$$

In particular, though $\mathscr{H}$ is a proper subclass of $\mathscr{T}^+$, there is a one-to-one correspondence between these two classes. In fact, if $f \in \mathscr{H}$ and $f^+ \in \mathscr{T}^+$, we put

$$f(z) = \frac{f^+(z) + z}{2},$$

i. e.,

$$f^+(z) = 2f(z) - z,$$

then the correspondence between

$$f^{+} \in \mathcal{J}^{+} \quad \text{and} \quad f \in \mathcal{H}$$

is bijective. Moreover, we see that there is an interrelation

$$\mathcal{L}^{\lambda} f^{+}(z) = 2\,\mathcal{L}^{\lambda} f(z) - z .$$

between them.

Under these circumstances, each result on either one of the classes $\mathcal{J}^{+}$ and $\mathcal{H}$ has its corresponding analogue on another class.

Now, with respect to the Hadamard convolution $*$, we have shown in § 3 that we get

$$\mathcal{L}^{\lambda} f = f * \mathcal{L}^{\lambda}\chi = \mathcal{L}^{\lambda}\chi * f ,$$

that is, the operator $\mathcal{L}^{\lambda}$ is represented also by $* \mathcal{L}^{\lambda}\chi$ or $\mathcal{L}^{\lambda} * \chi$ applied from the right or left, respectively, where χ denotes the definite function

$$\chi(z) = \frac{z}{1 - z}$$

which plays the role of identity with respect to the convolution within the class $\mathcal{J}$.

Now, we shall give the estimations of $\mathrm{Re}((\mathcal{L}^{\lambda} f(z))/z)$ for $f \in \mathcal{H}$ and $f \in \mathcal{J}^{+}$. Though this problem belongs to topics of a later category, the method for proof will illustrate the fact mentioned just above.

THEOREM 14.1. *For any* $f \in \mathscr{U}$, *the quantity defined by*

$$\mathcal{L}^{\lambda} f(z) = \int_I \frac{f(zt)}{t}\, d\sigma_{\lambda}(t)$$

satisfies for $r \in (0, 1)$ *and* $\lambda > 0$ *the estimation*

$$k_{\lambda}(r) \leq \operatorname{Re} \frac{\mathcal{L}^{\lambda} f(z)}{z} \leq K_{\lambda}(r)$$

along $\{|z| = r\}$, *where the bounds are given by*

$$\left.\begin{matrix} K_{\lambda}(r) \\ \\ k_{\lambda}(r) \end{matrix}\right\} = \int_I \frac{1}{1 \mp rt}\, d\sigma_{\lambda}(t) = \sum_{\nu=1}^{\infty} (\pm 1)^{\nu-1} \alpha_{\nu}^{\lambda} r^{\nu-1},$$

the α'*s being the moments with respect to* σ. *The equality sign in either inequality can occur only for*

$$f(z) = \bar{\varepsilon}\chi(\varepsilon z)$$

with $|\varepsilon| = 1$, *unless* $f(z) = z$.

Proof. From the integral representation of Herglotz [1], for the class $\mathcal{J}^{+}$ a structure formula for $f \in \mathscr{U}$ is readily obtained in the form

$$\frac{f(z)}{z} = \int_0^{2\pi} \frac{e^{i\theta}}{e^{i\theta} - z}\, d\tau(\theta)$$

$$= \int_0^{2\pi} \frac{e^{i\theta}\chi(e^{-i\theta} z)}{z}\, d\tau(\theta),$$

χ denoting the unit function of Hadamard convolution, where τ is a probability measure supported by the interval $[0, 2\pi)$. In view of this formula, we obtain

$$\frac{\mathcal{L}^{\lambda} f(z)}{z} = \int_0^{2\pi} \frac{\mathcal{L}^{\lambda}(e^{i\theta}\chi(e^{-i\theta}))}{z}\, d\tau(\theta),$$

whence follows the desired relation with

$$\left.\begin{matrix} K_{\lambda}(r) \\ k_{\lambda}(r) \end{matrix}\right\} = \left\{ \begin{matrix} \max \\ \min \\ |z|=r \end{matrix} \right. \operatorname{Re} \frac{\mathcal{L}^{\lambda}\chi(z)}{z} = \int_I \frac{1}{1 \mp rt}\, d\sigma_{\lambda}(t).$$

The assertion on the equality sign is readily verified. □

THEOREM 14.2. *For any* $f^{+} \in \mathcal{F}$ *we have for* $r \in (0, r)$ *and* $\lambda > 0$ *the estimation*

$$2k_{\lambda}(r) - 1 \leq \operatorname{Re}\frac{\mathcal{L}^{\lambda} f^{+}(z)}{z} \leq 2K_{\lambda}(r) - 1$$

along $\{|z| = r\}$. *The equality sign in either inequality can occur only for*

$$f^{+}(z) = 2\varepsilon\chi(\bar{\varepsilon}z) - z = z - \frac{\varepsilon + z}{\varepsilon - z}$$

with $|\varepsilon| = 1$, *unless* $f^{+}(z) = z$.

Proof. The procrdure similar to the proof of the preceding theorem applies by making use of Herglotz formula for $f^{+} \in \mathcal{F}^{+}$, namely,

$$\frac{f^+(z)}{z} = \int_0^{2\pi} \frac{e^{i\theta}+z}{e^{i\theta}-z}\, d\tau(\theta),$$

$$d\tau(\theta) \geqq 0, \qquad \int_0^{2\pi} d\tau(\theta) = 1.$$

Or indirectly, we may only take $f(z) = (f^+(z) + z)/2$ in the preceding theorem. □

In the particular case $\mathcal{L}^\lambda$ generated by $\sigma(t) = t$, the moment sequence is given by $\{\alpha_\nu\}_\nu = \{1/\nu\}$, and hence we get

$$\left.\begin{matrix} K_\lambda(r) \\ \\ k_\lambda(r) \end{matrix}\right\} = \sum_{\nu=1}^{\infty} \frac{(\pm 1)^{\nu-1}}{\nu^\lambda} r^{\nu-1}.$$

In general, $K_\lambda(r)$ is increasing while $k_\lambda(r)$ is decreasing, both with respect to $r \in (0, 1)$. In particular,

$$K_\lambda \equiv K_\lambda(1) = \sup_r K_\lambda(r) = \begin{cases} \infty & (0 < \lambda \leqq 1), \\ \zeta(\lambda) & (\lambda > 1), \end{cases}$$

$$k_\lambda \equiv k_\lambda(1) = \inf_r k_\lambda(r) = (1 - 2^{1-\lambda})\zeta(\lambda);$$

ζ denoting the Riemann zeta function.

Concerning the distortions on real part, we shall discuss more systematically in a later chapter.

§ 15. Univalent functions

In the present section, we shall first deal with the

whole class $\mathcal{S}$ of univalent functions normalized by

$$f(0) = f'(0) - 1 = 0;$$

For general theory, cf. Pommerenke [1].

We shall deal with its familiar subclasses $\mathcal{S}t$ and $\mathcal{K}$ also, though the results are quite elementary.

THEOREM 15.1. *If* $z\, d\mathcal{L}^{\lambda} f(z)/dz \in \mathcal{P}^{+}$, *then* $\mathcal{L}^{\mu} f \in \mathcal{S}$ *for* $\mu \geqq \lambda$.

Proof. The assumption $z\, d\mathcal{L}^{\lambda} f(z)/dz \in \mathcal{P}^{+}$ expresses, by definition, $\mathrm{Re}\, d\mathcal{L}^{\lambda} f(z)/dz > 0$, which is a sufficient condition for $\mathcal{L}^{\lambda} f \in \mathcal{S}$, in view of a theorem of Noshiro [1] and Wolff [2]; cf. also Warschawski [1]. If $\mu > \lambda$, we have

$$\frac{d}{dz} \mathcal{L}^{\mu} f(z) = \int_I \left(\frac{d}{dz} \mathcal{L}^{\lambda} f \right)(zt)\, d\sigma_{\mu-\lambda}(t)$$

whence follows that $\mathrm{Re}\, \mathcal{L}^{\lambda} f(z) > 0$ implies $\mathrm{Re}\, \mathcal{L}^{\mu} f(z) > 0$ and hence $\mathcal{L}^{\mu} f \in \mathcal{S}$. □

THEOREM 15.2. *If* $\mathcal{L}^{\lambda} f \in \mathcal{S}t$ *or* $\mathcal{L}^{\lambda} f \in \mathcal{K}$, *then* $\mathcal{L}^{\lambda+1} f \in \mathcal{S}t$ *or* $\mathcal{L}^{\lambda+1} f \in \mathcal{K}$, *respectively, provided* $\chi_1 \equiv \mathcal{L}\chi \in \mathcal{K}$.

Proof. Since $\mathcal{L}^{\lambda+1} f = \mathcal{L}^{\lambda} f * \mathcal{L}\chi = \mathcal{L}^{\lambda} f * \chi_1$, we only have to remember a theorem of Ruscheweyh and Sheil-Small [1] giving the affirmative answer to a conjecture of Pólya-Schoenberg [1]. □

In the distinguished case, we can derive several similar

theorems, of which a few will be stated below.

THEOREM 15.3. *Let* $\mathcal{L}$ *be the operator generated by* $\sigma(t) = t$. (i) *If* $\mathcal{L}^{\lambda} f \in \mathcal{F}^{+}$, *then* $\mathcal{L}^{\mu} f \in \mathcal{S}$ *for* $\mu \geqq \lambda + 1$. (ii) *If* $z \mathcal{L}^{\lambda+1} f / \mathcal{L}^{\lambda} f \in \mathcal{F}^{+}$, *then* $\mathcal{L}^{\lambda+1} f \in \mathcal{S}t$. (iii) *If* $\mathcal{L}^{\lambda} f \in \mathcal{S}t$, *then* $\mathcal{L}^{\lambda+1} f \in \mathcal{K}$.

Proof. (i) Since $\mathcal{L}$ coincides with the integration with respect to log z, we have

$$z \frac{d}{dz} \mathcal{L}^{\lambda+1} f(z) = \mathcal{L}^{\lambda} f(z)$$

in the distinguished case. Consequently, $\mathcal{L}^{\lambda} f \in \mathcal{F}^{+}$ implies $z(d/dz)\mathcal{L}^{\lambda+1} f(z) \in \mathcal{F}^{+}$, whence follows, in view of general Theorem 14.3, $\mathcal{L}^{\mu} f \in \mathcal{S}$ for $\mu \geqq \lambda + 1$.

(ii) If $z \mathcal{L}^{\lambda+1} f / \mathcal{L}^{\lambda} f \in \mathcal{F}^{+}$, we have by definition Re $(\mathcal{L}^{\lambda+1} f / \mathcal{L}^{\lambda} f) > 0$ and hence

$$\operatorname{Re} \frac{(d/dz)\mathcal{L}^{\lambda+1} f(z)}{\mathcal{L}^{\lambda+1} f(z)} = \operatorname{Re} \frac{\mathcal{L}^{\lambda} f(z)}{\mathcal{L}^{\lambda+1} f(z)} > 0$$

which is the condition for $\mathcal{L}^{\lambda+1} f \in \mathcal{S}t$.

(iii) If $\mathcal{L}^{\lambda} f \in \mathcal{S}t$, then by a theorem of Alexander [1]

$$\mathcal{L}^{\lambda+1} f(z) = \int_0^z \frac{\mathcal{L}^{\lambda} f(\zeta)}{\zeta} d\zeta \in \mathcal{K}. \qquad \square$$

Now, we have already shown in Theorem 4.3 the limit behavior of $\mathcal{L}^{\lambda} f(z)$ as λ tends to ∞; namely, it tends to z as $\lambda \to \infty$ to z uniformly in the wider sense in E, unless $\mathcal{L} f = f$.

In general, there are of course several sequences in $\mathcal{F}$ which tend to univalent functions uniformly in the wider sense but it is possible that no member of such a sequence is univalent in E. Nevertheless, the family $\{\mathcal{L}^{\lambda} f\}$ shows a fairly tractable behavior with respect to λ, and this fact permits us, for instance, to state the following theorem:

THEOREM 15.4. *Let* $f(z) = \sum_{\nu=1}^{\infty} c_{\nu} z^{\nu} \in \mathcal{F}$ *satisfy the condition that there exists a constant* N *for which* $c_{\nu} = O(\nu^{N})$ *as* $\nu \to \infty$. *Then in the distinguished case of* $\mathcal{L}^{\lambda}$ *generated by* $\sigma(t) = t$, *we have* $\mathcal{L}^{\lambda} f \in St$ *provided* λ *is large enough.*

Proof. We may suppose that there exists a positive integer M ($\geqq 3$) satisfying

$$|c_{\nu}| \leqq M\nu^{\lambda-4} \qquad (\nu = 2, 3, \ldots)$$

for $\lambda \geqq N + 4$. For such a value of M, we can find a value $\Lambda \geqq N + 4$ such that

$$\sum_{\nu=2}^{M-1} \frac{|c_{\nu}|}{\nu^{\Lambda-1}} < 2 - \frac{\pi^2}{6}.$$

Then we obtain for $\lambda \geqq \Lambda$

$$\sum_{\nu=2}^{\infty} \frac{|c_{\nu}|}{\nu^{\lambda-1}} \leqq \sum_{\nu=2}^{M-1} \frac{|c_{\nu}|}{\nu^{\lambda-1}} + \sum_{\nu=M}^{\infty} \frac{M}{\nu^{3}}$$

$$< 2 - \frac{\pi^2}{6} + \sum_{\nu=M}^{\infty} \frac{1}{\nu^2} < 1.$$

The last inequality shows that a condition sufficient for the starlikeness of

$$\mathcal{L}^{\lambda} f(z) = z + \sum_{\nu=2}^{\infty} \frac{c_\nu}{\nu^{\lambda}} z^{\nu}$$

is satisfied. □

The assumption in the Theorem 15.4 just proved provides a fairly wide subclass of the original class $\mathcal{F}$ which is restricted by $c_1 = 1$ and $\limsup_{\nu\to\infty} |c_\nu|^{1/\nu} \leq 1$. As an example, we state here one of its corollaries:

COROLLARY 15.1. *In the distinguished case of* $\mathcal{L}^{\lambda}$, *if there exists a* κ *such that* $\inf_{z \in E} \mathrm{Re}(\mathcal{L}^{\kappa} f(z)/z) > -\infty$, *then we have* $\mathcal{L}^{\lambda} f \in St$, *provided* λ *is large enough.*

Proof. Suppose, by definition,

$$\mathrm{Re} \frac{f_\kappa(z)}{z} \geq h_\kappa > -\infty \quad \text{in } E.$$

Then we see that

$$\frac{f_\kappa(z) - h_\kappa z}{1 - h_\kappa} \equiv z + \sum_{\nu=2}^{\infty} d_\nu z^\nu \in \mathcal{J}^+$$

and hence $|d_\nu| \leqq 2$. Hence, the Taylor coefficients of f_κ are bounded so that in view of Theorem 15.4 $\mathcal{L}^\lambda f \equiv \mathcal{L}^{\lambda-\kappa}\mathcal{L}^\kappa f \in \mathcal{S}t$ provided λ is large enough. □

In conclusion, we give a theorem on the starlikeness of $\mathcal{L}^\lambda f$, of which the proof depends on numerical computation.

THEOREM 15.5. *In the distinguished case generated by* $\sigma(t) = t$, *if* $f \in \mathcal{S}$, *then* $\mathcal{L}^\lambda f \in \mathcal{S}t$ *at least for* $\lambda \geqq \lambda_0$ *where* λ_0 *is a unique root of the transcendental equation*

$$\zeta(\lambda - 2) = 2$$

contained in the interval $(3, 4)$.

Proof. Let

$$f(z) = z + \sum_{\nu=2}^{\infty} c_\nu z^\nu \in \mathcal{S}.$$

Since the conjecture of Bieberbach [1]: $|c_\nu| \leqq \nu$ has been affirmatively solved by de Branges [1], we get

$$\sum_{\nu=2}^{\infty} \frac{|c_\nu|}{\nu^{\lambda-1}} \leqq \sum_{\nu=2}^{\infty} \frac{1}{\nu^{\lambda-2}} = \zeta(\lambda - 2) - 1.$$

Hence we see that a condition

$$\sum_{\nu=2}^{\infty} \frac{|c_\nu|}{\nu^{\lambda-1}} \leqq 1$$

sufficient for $\mathcal{L}^\lambda f \in St$ is satisfied for $\lambda \geqq \lambda_0$ with a unique root λ_0 of the transcendental equation

$$\zeta(\lambda - 2) = 2. \qquad \square$$

In connection with Theorem 15.5, it seems plausible to conjecture that if $f \in \mathcal{S}$, then $\mathcal{L}^\lambda f \in \mathcal{S}$ at least for $\lambda \geqq 1$, and if $f \in \mathcal{K}$ (or, more generally, $f \in St$), then $\mathcal{L}^\lambda f \in \mathcal{K}$ at least for $\lambda \geqq 1$, both in the distinguished case generated by $\sigma(t) = t$.

In fact, Biernacki [1] has advanced an affiirmative proof of the former conjecture for $\lambda = 1$. However, the proof contains unfortunately a flaw. Though the indicated flaw did not show by itself whether Biernacki's theorem could be saved. Nevertheless, Krzyż and Lewandowski [1] have really constructed a counter-example which shows that there exists a function $f \in \mathcal{S}$ such that $\mathcal{L} f$ is not univalent in E.

Let $SP(\alpha)$ with $\alpha \in [-\pi/2, \pi/2]$ be the class of functions $f \in \mathcal{F}$ such that a real constant α may be chosen so that

$$\operatorname{Re} e^{i\alpha} z \frac{f'(z)}{f(z)} > 0$$

throughout E. In order to state a counter-example shown by Krzyż and Lewandowski, we need a theorem obtained by Špaček

[1] which will be referred to as a Lemma in the following lines:

LEMMA 15.1. $\mathcal{SP}(\alpha)$ is a subclass of $\mathcal{S}$.

Proof (cf. also Goodman [1], Vol. 1, p. 149). If $f \in \mathcal{SP}(\alpha)$ did not belong to $\mathcal{S}$, then there is some $r \in (0, 1)$ such that f takes the same value at two distinct points on $C_r \equiv \{|z| = r < 1\}$. Suppose that

$$f(z_1) = f(z_2), \quad z_\nu = re^{i\theta_\nu} \quad (\nu = 1, 2)$$

with $\theta_1 \not\equiv \theta_2 \pmod{2\pi}$. Since f has only one simple zero in E at the origin, the rotation number of $\arg f$ along C_r is equal to unity. Hence, θ_1 and θ_2 can be chosen so that $0 \leq \theta_1 < \theta_2 < 2\pi$ and

$$\Gamma \equiv \{f(z = re^{i\theta}), \ \theta_1 \leq \theta \leq \theta_2\}$$

is a simple closed curve which does not enclose the origin. Hence $\log f(z)$ is single-valued along Γ and $\log f(z_1) = \log f(z_2)$. Consequently, we have

$$0 = \int_\Gamma \frac{d}{dz} \log f(z)\, dz = i \int_{\theta_1}^{\theta_2} z \frac{f'(z)}{f(z)}\, d\theta.$$

On the other hand, by the assumption $f \in \mathcal{SP}(\alpha)$, the value of the integrand must always lie in a half-plane bounded by a straight line through the origin. Hence we have

$$\log f(z_2) - \log f(z_1) = \int_\Gamma \frac{f'(z)}{f(z)}\, dz$$

$$[z = re^{i\theta}]$$

$$= e^{-i\alpha} \int_{\theta_1}^{\theta_2} e^{i\alpha} z \frac{f'(z)}{f(z)}\, d\theta,$$

The last member cannot vanish in view of $f \in SP(\alpha)$, what is a contradiction. □

Consider now the function

$$f_0(z) = z \exp((i - 1) \log(1 - iz)),$$

log denoting its principal branch. Since

$$\operatorname{Re} e^{i\pi/4} z \frac{f_0'(z)}{f_0(z)} = \frac{1}{2^{1/2}} \operatorname{Re} \frac{(1 + i)(1 + z)}{1 - iz}$$

$$= \frac{1}{2^{1/2}} \frac{1 - |z|^2}{|1 - iz|^2} > 0$$

in E, $f_0 \in SP(\pi/4)$. On the other hand,

$$\mathcal{L} f_0(z) = \int_0^z \frac{f_0(\zeta)}{\zeta}\, d\zeta = \exp(i \log(1 - iz)) - 1.$$

Putting now $z_1 = i(e^{2\pi} - 1)/(e^{2\pi} + 1)$ and $z_2 = -z_1 \in E$, we see that

$$\mathcal{L} f_0(z_1) = \exp\left(i \log\left(1 + \frac{e^{2\pi} - 1}{e^{2\pi} + 1}\right)\right) - 1$$

$$= \exp\left(i \log\left(1 - \frac{e^{2\pi} - 1}{e^{2\pi} + 1}\right)\right) - 1 = \mathcal{L} f_0(z_2),$$

which shows that $\mathcal{L} f(z)$ is not univalent in E.

We now observe again the class $\mathcal{S}$ of functions f univalent in E. The Bieberbach conjecture asserting

$$(15.1) \qquad |f^{(n)}(0)| \leqq n!\, n \qquad (n = 2, 3, \ldots)$$

has been affirmatively solved by de Branges. And the equality sign appears for every n if and only if f coincides with a Koebe function $\bar{\varepsilon} F(\varepsilon z)$, where we put

$$F(z) = \frac{z}{(1 - z)^2}.$$

Concerning a history of the Bieberbach conjecture, cf. Pommerenke [1], pp. 24 et seq. On the other hand, Landau [1] has shown (see Lemma 15.2 below) that, if (15.1) would hold, then the following distortion inequality is valid:

$$(15.2) \quad |f^{(n)}(z)| \leqq n! \frac{n + |z|}{(1 - |z|)^{n+1}} = F^{(n)}(|z|)$$

with the same extremal function $\bar{\varepsilon} F(\varepsilon z)$. Now, (15.1) having been verified for $\mathcal{S}$, the distortion inequality (15.2) always holds also. We show the following theorem; cf. Komatu [7]:

THEOREM 15.6. *Let f belong to $\mathcal{S}$ and λ be non-negative.*

Then, every $\mathcal{L}^{\lambda} f$ *satisfies the distortion inequality*

$$(15.3)\quad |\mathcal{L}^{\lambda} f^{(n)}(z)| \leqq \mathcal{L}^{\lambda} F^{(n)}(|z|) \quad (n = 2, 3, \ldots)$$

where F *denotes the Koebe function. The extremal function is only given by* $\bar{\varepsilon} F(\varepsilon z)$.

Proof. We set for the sake of brevity

$$\sigma_{\lambda}(t) = \int_0^t \left(\log \frac{1}{\tau}\right) d\tau$$

and $f_{\lambda}(z) = \mathcal{L}^{\lambda} f(z)$. Then we have

$$f_{\lambda}^{(n)}(z) \equiv \mathcal{L}^{\lambda} f^{(n)}(z)$$

$$= \int_I t^{n-1} f^{(n)}(zt)\, d\sigma_{\lambda}(t).$$

Since f satisfies (15.2), we have

$$|f^{(n)}(zt)| \leqq F^{(n)}(|z|t)$$

and hence obtain

$$|f_{\lambda}^{(n)}(z)| \leqq \int_I t^{n-1} F^{(n)}(|z|t)\, d\sigma_{\lambda}(t)$$

$$= F_{\lambda}^{(n)}(|z|).$$

The extremality assertion follows from that of (15.2). □

On the other hand, it has been shown before that $\mathcal{L}$ is nothing but the integration with respect to log z, $\mathcal{L}^{\lambda}$ can be analytically prolongable by means of

$$f_\lambda(z) = \left(\frac{d}{d\log z}\right)^m f_{\lambda+m}(z)$$

with any positive integer m.

Then, it will be shown that the assertion in Theorem 15. 6 remains true without the restriction $\lambda \geqq 0$:

THEOREM 15.7. *The assertion in Theorem* 15.6 *is valid for any real value of* λ.

Proof. We only have to verify the case $\lambda < 0$. We may take, for instance, $p = -[\lambda]$. Then, if we perform, in the right-hand member of the equation

$$f_\lambda(z) = \left(\frac{d}{d\log z}\right)^p f_{\lambda+p}(z),$$

the differentiation with respect to log z, it causes a linear combination of $z^j f_{\lambda+p}^{(j)}(z)$ $(j = 1, \ldots, p)$ with positive coefficients alone. Hence, $f_\lambda^{(n)}(z)$ is a fixed linear combination of

$$z^\nu f_{\lambda+p}^{(n-\nu)}(z) \quad (\nu = 0, 1, \ldots, n)$$

with positive coefficients alone. Accordingly, if we put

$$f_\lambda^{(n)}(z) = \sum_{\nu=1}^{n} b_\nu z^\nu f_{\lambda+p}^{(n-\nu)}(z), \qquad b_\nu > 0,$$

then we get, in view of Theorem 15.6 and by the structure of production of the related linear combination, the estimation

$$|f_\lambda^{(n)}(z)| \leq \sum_{\nu=0}^{n} b_\nu |z|^\nu F_{\lambda+p}^{(n-\nu)}(|z|)$$

$$= \frac{d^n}{d|z|^n}\left(\frac{d}{d\log|z|}\right)^p F_{\lambda+p}(|z|)$$

$$= F_\lambda^{(n)}(|z|).$$

The extremality assertion follows also from the above development of the proof. □

Though the method of proof has been based on the integral representation for $\mathcal{I}^\lambda$ until now, it is also possible to make use of power series expansion. In fact, if the Taylor expansion of a function $f \in \mathcal{I}$ is

$$(15.4) \qquad f(z) = \sum_{\nu=1}^{\infty} c_\nu z^\nu \quad \text{with} \quad c_1 = 1,$$

then we have

$$(15.5) \qquad f_\lambda(z) \equiv \mathcal{L}^\lambda f(z) = \sum_{\nu=1}^{\infty} \alpha_\nu^\lambda c_\nu z^\nu$$

with

$$\alpha_\nu = \int_I t^{\nu-1}\, d\sigma(t) \qquad (\nu = 1,\ 2,\ \ldots).$$

We begin with the Landau's assertion as a lemma:

LEMMA 15.2. *If* $f \in \mathcal{J}$ *satisfies* (15.1), *then the distortion inequality* (15.2) *holds for* f *in* E.

Proof. Based on the Taylor expansion of f, we have

$$|f^{(n)}(z)| = \left| \frac{d^n}{dz^n} \sum_{\nu=1}^{\infty} \frac{f^{(\nu)}(0)}{\nu\,!} z^\nu \right|$$

$$= \left| \sum_{\nu=n}^{\infty} \frac{\nu\,!}{(\nu - n)!} \frac{f^{(\nu)}(0)}{\nu\,!} z^{\nu-n} \right|$$

$$\leqq \sum_{\nu=n}^{\infty} \frac{\nu\,!}{(\nu - n)!} \nu\, |z|^{\nu-n} = F^{(n)}(|z|). \qquad \square$$

THEOREM 15.8. *Under the same assumption as in Lemma* 15.2, *the distortion inequality* (15.3) *holds for every integer* n *and every real* λ. *The extremal function is the same as in Theorem* 15.6.

Proof. By making use of Lemma 15.2, the proof of Theorem remains valid literally. Or alternatively, by using the series form (15.4), we have more directly from (15.5)

$$|f_\lambda^{(n)}(z)| = \left| \sum_{\nu=n}^{\infty} \frac{\nu\,!}{(\nu - n)!} \alpha_\nu^{\lambda} c_\nu z^{\nu-n} \right|$$

$$\leqq \sum_{\nu=n}^{\infty} \frac{\nu\,!}{(\nu - n)!} \alpha_\nu^{\lambda}\, \nu\, |z|^{\nu-n}$$

$$= F_\lambda^{(n)}(|z|). \qquad \square$$

Here we notice by the way that $f_\lambda^{(n)}(z)$ with $\lambda > n + 2$ is uniformly bounded with respect to $z \in E$, provided only

$$c_\nu = \frac{f^{(\nu)}(0)}{\nu\,!} = O(\nu).$$

In fact, $|f_\lambda^{(n)}(z)|$ has then an upper bound

$$F_\lambda^{(n)}(1) = \sum_{\nu=n}^{\infty} \frac{\nu\,!}{(\nu - n)!} \frac{\nu}{\nu^\lambda}$$

except a constant factor independent of z, where the term in the summation satisfies

$$\frac{\nu\,!}{(\nu - n)!} \frac{\nu}{\nu^\lambda} \leqq \frac{1}{\nu^{\lambda-n-1}}, \qquad \lambda - n - 1 > 1$$

and hence the series itself does converge.

In conclusion, it is noticed that in the distinguished case generated by $\sigma(t) = t$, the bound $F_\lambda^{(n)}$ often appeared for Koebe function F is expressible in the scope of integral of elementary functions. In fact, we obtain

$$F_\lambda(r) = \frac{1}{\Gamma(\lambda)} \int_I \frac{F(rt)}{t} \left(\log \frac{1}{t}\right)^{\lambda-1} dt$$

$$= \frac{1}{\Gamma(\lambda)} \int_0^r \frac{F(\tau)}{\tau} \left(\log \frac{r}{\tau}\right)^{\lambda-1} d\tau$$

and hence

$$F_\lambda'(r) = \frac{F_{\lambda-1}(r)}{r} .$$

In particular, this bound can be given by elementary functions explicitly. For instance,

$$F(r) = F_0(r) = \frac{r}{(1-r)^2}, \qquad F_1(r) = \frac{r}{1-r},$$

$$F_2(r) = \log\frac{1}{1-r}, \qquad F_3(r) = \frac{1}{r} \log\frac{1}{1-r} .$$

§ 16. Particular subclasses

In the preceding sections, we have referred to some functions starlike with respect to the origin and convex as familiar subclasses of univalent functions.

However, besides these subclasses there are several particular ones which were in detail studied from various viewpoints. For instance, the order of starlike functions has been introduced by Holland and Thomas [1], Libera [1], Padmanabhhan [1] and Robertson [1] etc.

According to Padmanabhan, a function $f \in \mathcal{S}$ is said to be starlike of order α in E if

$$\left| \left(\frac{z f'(z)}{f(z)} - 1 \right) \Big/ \left(\frac{z f'(z)}{f(z)} + 1 \right) \right| < \alpha$$

for $\alpha \in (0, 1]$. We denote by $\mathcal{S}t\,(\alpha)$ the class of starlike functions of order α.

Mogra [1] has derived a coefficient theorem on $\mathcal{S}t\,(\alpha)$ and a sufficient condition for

$$f(z) = z + \sum_{\nu=2}^{\infty} c_\nu z^\nu \tag{16.1}$$

to be in $\mathcal{S}t\,(\alpha)$, which will be stated as in the following lemmas:

LEMMA 16.1. *If* f *defined by* (16.1) *is in the class* $\mathcal{S}t\,(\alpha)$ $(0 < \alpha \le 1)$, *then for* $\alpha = 1$,

$$|c_\nu| \le \nu \qquad (\nu \ge 2) \tag{16.2}$$

while for $0 < \alpha < 1$,

$$|c_\nu| \le \nu\, \alpha^{\nu-1} \qquad \textit{for } 2 \le \nu \le N \tag{16.3}$$

and

(16.4) $$|c_\nu| \leq \frac{1}{\nu - 1} \alpha^N N(N+1) \qquad \textit{for } \nu > N$$

where $N = [(1+\alpha)/(1-\alpha)]$; *here* $[\ ]$ *denoting the Gauss' symbol.* □

Proof. The definition of $F \in St$ asserts that $zF'/F(z)$ is is contained within the circle $\{|(w-1)/(w+1)| < \alpha\}$. Now,

$$w = \varphi(z) \equiv \frac{1+\alpha z}{1-\alpha z}$$

maps E onto this circle univalently and $f(0) = 1$. Hence, $zF'(z)/F(z)$ is subordinate to $\varphi(z)$. Consequently, F is subordinate to

$$\phi(z) \equiv \exp \int_0^z \frac{1}{\zeta} \varphi(\zeta)\, d\zeta = \frac{z}{(1-\alpha z)^2}$$

$$= \sum_{\nu=1}^{\infty} \nu\, \alpha^{\nu-1},$$

and the result follows. Cf. remark below. □

REMARK. Though Mogra stated in his paper that the proof of the theorem is similar to that of Clunie [1] and hence is omitted, we have introduced here a rather simple proof. It is noted that it is unnecessary to distinguish between the cases of $\alpha = 1$ and $0 < \alpha < 1$, since in the latter case a circle reduces only to a half-plane. Morover, the inequality of (16.4) is involved in that of (16.3).

LEMMA 16.2. *If f defined by* (16.1) *satisfies*

$$\sum_{\nu=2}^{\infty} \left(\left(\frac{1+\alpha}{2\alpha} \right) \nu - \frac{1-\alpha}{2\alpha} \right) |c_\nu| \leqq 1, \tag{16.5}$$

then $f \in \mathcal{S}t\,(\alpha)$.

Proof. For $|z| = r < 1$, we have

$$|zf'(z) - f(z)| - \alpha |zf'(z) + f(z)|$$

$$= \left| \sum_{\nu=2}^{\infty} (\nu - 1) c_\nu z^\nu \right| - \alpha \left| 2z + \sum_{\nu=2}^{\infty} (\nu + 1) c_\nu z^\nu \right|$$

$$\leqq \sum_{\nu=2}^{\infty} (\nu - 1) |c_\nu| r^\nu - \alpha \left(2r - \sum_{\nu=2}^{\infty} (\nu + 1) |c_\nu| r^\nu \right)$$

$$< \left(\sum_{\nu=2}^{\infty} (\nu - 1) |c_\nu| - 2\alpha + \sum_{\nu=2}^{\infty} \alpha (\nu + 1) |c_\nu| \right) r$$

$$= \left(\sum_{\nu=2}^{\infty} ((1 + \alpha)\nu - (1 - \alpha)) \vert c_\nu - 2\alpha \right) r$$

$$\leqq 0$$

by the assumption. Hence it follows that $f \in \mathcal{S}t\,(\alpha)$. □

REMARK. It is noted that $f(z) = z - (2\alpha)/((1 + \alpha)\nu - (1 - \alpha)) z^\nu$ is extremal with respect to the Theorem, since

$$\left| \left(z \frac{f'(z)}{f(z)} - 1 \right) \Big/ \left(z \frac{f'(z)}{f(z)} + 1 \right) \right| = \alpha$$

for $z = 1$, $0 < \alpha \leqq 1$ and $\nu = 2, 3, \ldots$.

It is also observed that the converse to the Theorem is false. In fact, $f(z) = z/(1 - \alpha z)^2 \in \mathcal{S}t(\alpha)$, while $c_\nu = \nu\alpha^{\nu-1}$ and hence for $0 < \alpha \leqq 1$

$$\sum_{\nu=2}^{\infty} \left(\frac{1 + \alpha}{2\alpha} \nu - \frac{1 - \alpha}{2\alpha} \right) |c_\nu| > 1.$$

By making use of these Lemmas, Owa [1] has derived the following two theorems; bounds of λ_* being somewhat improved.

THEOREM 16.1. *Let f defined by* (16.1) *belong to the class* $\mathcal{S}t\,(1/3)$. *Then $\mathcal{L}^\lambda f$ belongs to the same class at least for $\lambda \geqq \lambda_*$ where λ_* is a certain number less than* 43/16.

Proof. Putting $\alpha = 1/3$, we have $N = 2$ in Lemma 15.1. Since

$$\mathcal{L}^\lambda f(z) = z + \sum_{\nu=2}^{\infty} \frac{c_\nu}{\nu^\lambda} z^\nu,$$

the sufficient condition for $\mathcal{L}^\lambda f \in \mathcal{S}t\,(1/3)$ may be verified with the aid of Lemma 16.2 as in the following manner:

$$\sum_{\nu=2}^{\infty} (2\nu - 1) \frac{|c_\nu|}{\nu^\lambda} = \frac{3\,|c_2|}{2^\lambda} + \sum_{\nu=3}^{\infty} (2\nu - 1) \frac{|c_\nu|}{\nu^\lambda}$$

$$\leqq \frac{1}{2^{\lambda-1}} + \frac{2}{3}\sum_{\nu=3}^{\infty} \frac{2\nu - 1}{\nu^{\lambda}(\nu - 1)}$$

$$< \frac{1}{2^{\lambda-1}} + \frac{2}{3}\sum_{\nu=3}^{\infty} \frac{1}{\nu^{\lambda-1}}$$

$$< \frac{1}{2^{\lambda-1}} + \frac{2}{3}\int_{2}^{\infty} \frac{dx}{x^{\lambda-1}}$$

$$= \left(1 + \frac{4}{3(\lambda - 2)}\right)\frac{1}{2^{\lambda-1}},$$

and the last member of this estimation is reaslly less than the unity, provided $\lambda \geqq 43/16$. □

THEOREM 16. 2. *Let* f *defined by* (16.1) *belong to the class* $\mathcal{S}t\,(1/2)$. *Then* $\mathcal{L}^{\lambda} f$ *belongs to the same class at least for* $\lambda \geqq \lambda_*$ *where* λ_* *is a certain number less* than 11/4.

Proof. Put $\alpha = 1/2$. Then in Lemma 16.1 we have $N = 3$. By making use of Lemmas 16.1, the sufficient condition in Lemma 16.2 can be verified in a quite similar manner as in the preceding Theorem 16.1. That is, we have

$$\sum_{\nu=2}^{\infty} \frac{3\nu - 1}{2}\frac{|c_{\nu}|}{\nu^{\lambda}} = \frac{5|c_2|}{2^{\lambda+1}} + \frac{4|c_3|}{3^{\lambda}} + \frac{1}{2}\sum_{\nu=4}^{\infty}(3\nu - 1)\frac{|c_{\nu}|}{\nu^{\lambda}}$$

$$= \frac{5}{2^{\lambda+1}} + \frac{1}{3^{\lambda-1}} + \frac{3}{4}\int_{3}^{\infty}\frac{dx}{x^{\lambda-1}}$$

$$= \frac{5}{2^{\lambda+1}} + \left(1 + \frac{9}{4(\lambda - 2)} \right) \frac{1}{3^{\lambda-1}} .$$

The last member of this estimation is really less than the unity provided $\lambda \leqq 11/4$. □

After showing Theorem 16.1, Owa has paused the following problem:

Let f defined by (16.1) be in the class $\mathcal{St}\,(\alpha)$ with $0 < \alpha \leqq 1$. Then, does $\mathcal{L}^{\lambda} f$ belong to the same class at least for $\lambda \geqq \lambda_{*}$ with a certain number λ_{*} less than 3 ?

On the other hand, Owa [2] has observed some special classes and dealt with a conjecture with respect to the problem to determine the lower bound for λ_{*} which was presented originally in the general class by the present author; cf. Komatu [5].

Now, in order to deal with other subclasses, consider the functions f holomorphic in E whose Taylor expansions have real coefficients alone and moreover are of the form

$$(16.6) \qquad f(z) = z - \sum_{\nu=2}^{\infty} a_{\nu} z^{\nu} \qquad (a_{\nu} \geqq 0).$$

Let $\mathcal{T}$, $\mathcal{T}^{*}(\alpha)$, and $\mathcal{C}(\alpha)$ denote the classes consisting of such functions (16.7) which are univalent, starlike of order α with respect to the origin, and convex of order α, respectively, α being a parameter satisfying $0 \leqq \alpha < 1$.

Let $P^*(\alpha, \beta)$, $S^*(\alpha, \beta)$, and $C^*(\alpha, \beta)$ denote the classes consisting of functions f in (15.7) such that

$$\left| \frac{f'(z) - 1}{f'(z) + 1 - 2\alpha} \right| < \beta,$$

$$\left| \frac{zf'(z)/f(z) - 1}{zf'(z)/f(z) + 1 - 2\alpha} \right| < \beta,$$

and

$$zf' \in S^*(\alpha, \beta),$$

respectively, where α and β are parameters satisfying $0 \leqq \alpha < 1$, $0 < \beta \leqq 1$.

We state two lemmas about these classes, of which the former is due to Silverman [1] while the latter is due to Gupta and Jain [1, 2].

LEMMA 16.3. *A function f defined by* (16.7) *is in T, $T^*(\alpha)$ or $C(\alpha)$ if and only if*

$$\sum_{\nu=2}^{\infty} \nu a_\nu \leqq 1,$$

$$\sum_{\nu=2}^{\infty} (\nu - \alpha) a_\nu \leqq 1 - \alpha,$$

or

$$\sum_{\nu=2}^{\infty} \nu(\nu - \alpha) a_\nu \leqq 1 - \alpha,$$

respectively.

Proof. We shall sketch a proof of this Lemma. We first show that $f \in T^*(\alpha)$ if and only if $\sum_{\nu=2}^{\infty}(\nu - \alpha) a_\nu \leqq 1 - \alpha$. For that purpose, suppose this condition is satisfied. Then,

$$\sum_{\nu=2}^{\infty} (n - 1) a_\nu \leqq (1 - \alpha)\left(1 - \sum_{\nu=2}^{\infty} a_\nu\right)$$

and hence

$$\left| z\frac{f'(z)}{f(z)} - 1\right| = \left| \frac{\sum_{\nu=2}^{\infty} (\nu - 1) a_\nu z^\nu}{z + \sum_{\nu=2}^{\infty} a_\nu z^\nu} \right|$$

$$\leqq \frac{\sum_{\nu=2}^{\infty} (\nu - 1) a_\nu z^{\nu-1}}{1 - \sum_{\nu=2}^{\infty} a_\nu z^{\nu-1}}$$

$$\leqq \frac{\sum_{\nu=2}^{\infty} (\nu - 1) a_\nu}{1 - \sum_{\nu=2}^{\infty} a_\nu} \leqq 1 - \alpha.$$

The last inequality implies $\operatorname{Re}(zf'/f) > \alpha$, i. e., $f \in \mathcal{T}^*(\alpha)$. Conversely, if $\operatorname{Re}(zf'/f) > \alpha$ in E, we obtain

$$\operatorname{Re} z\frac{f'(z)}{f(z)} = \operatorname{Re}\frac{z - \sum_{\nu=2}^{\infty} \nu a_\nu z^\nu}{z - \sum_{\nu=2}^{\infty} a_\nu z^\nu} > \alpha.$$

Choose z on the real axis so that zf'/f is real. Let $z \to 1 - 0$ through real values, we obtain

$$1 - \sum_{\nu=2}^{\infty} \nu a_\nu \leqq \alpha\left(1 - \sum_{\nu=2}^{\infty} a_\nu\right), \text{ i. e., } \sum_{\nu=2}^{\infty} (\nu - \alpha) a_\nu \leqq 1 - \alpha.$$

For the case of $\mathcal{C}(\alpha)$, in view of the theorem of Alexander [1], we have only to replace a_ν by νa_ν. For the case of $\mathcal{T}$, suppose $\sum_{\nu=2}^{\infty} \nu a_\nu = 1 + \delta > 1$. Then, there exists an integer N such that $\sum_{\nu=2}^{N} \nu a_\nu > 1 + \delta/2$. For $z \in (1/(1 + \delta/2))^{1/(N-1)}$ we have

$$\begin{aligned} f'(z) &\leqq 1 - \sum_{\nu=2}^{N} \nu a_\nu z^{\nu-1} \\ &\leqq 1 - z^{N-1} \sum_{\nu=2}^{N} \nu a_\nu < 1 - \left(1 + \frac{\delta}{2}\right) z^{N-1} < 0. \end{aligned}$$

Since $f'(0) > 0$, there exists a real number $z_0 \in (0, 1)$ for which $f'(z_0) = 0$, and hence f does not belong to $\mathcal{T}$. □

Moreover, we have seen that $\mathcal{T} = \mathcal{T}^*(0)$.

LEMMA 16.4. *A function* f *defined by* (15.7) *is in the class* $P^*(\alpha, \beta)$, $S^*(\alpha, \beta)$ *or* $C^*(\alpha, \beta)$ *if and only if*

$$\sum_{\nu=2}^{\infty} \nu(1 + \mu) a_\nu \leqq 2\beta(1 - \alpha),$$

$$\sum_{\nu=2}^{\infty} (\nu - 1 + \beta(\nu + 1 - 2\alpha)) a_\nu \leqq 2\beta(1 - \alpha),$$

or

$$\sum_{\nu=2}^{\infty} \nu(\nu - 1 + \beta(\nu + 1 - 2\alpha)) a_\nu \leqq 2\beta(1 - \alpha),$$

respectively.

Proof. If the condition on P^* is satisfied, then for $|z| = 1$

$$|f'(z) - 1| - \beta|f'(z) + 1 - 2\alpha|$$

$$= \left|- \sum_{\nu=2}^{\infty} \nu a_\nu z^{\nu-1}\right| - \beta\left|2(1 - \alpha) \sum_{\nu=2}^{\infty} \nu a_\nu z^{\nu-1}\right|$$

$$\leqq \sum_{\nu=2}^{\infty} \nu(1 + \beta) a_\nu - 2\beta(1 - \alpha) \leqq 0.$$

Hence, by maximum modulus theorem, $f \in P^*(\alpha, \beta)$. For the converse, suppose that

$$\left|\frac{f'(z) - 1}{f'(z) + 1 - 2\alpha}\right|$$

$$= \left|\left(- \sum_{\nu=2}^{\infty} \nu a_\nu z^{\nu-1}\right)\Big/\left(2(1 - \alpha) - \sum_{\nu=2}^{\infty} \nu a_\nu z^{\nu-1}\right)\right| < \beta,$$

then, we have

$$\mathrm{Re}\left(\left(\sum_{\nu=2}^{\infty} \nu a_\nu z^{\nu-1}\right)\Big/\left(2(1 - \alpha) - \sum_{\nu=2}^{\infty} \nu a_\nu z^{\nu-1}\right)\right) < \beta.$$

Choose values of z on the real axis so that $f'(z)$ is real. Then, letting $z \to 1 - 0$ through real values, we obtain

$$\sum_{\nu=2}^{\infty} \nu a_\nu \leqq 2\beta(1 - \alpha) - \beta \sum_{\nu=2}^{\infty} \nu a_\nu,$$

which yields the desired result. The case of $\mathcal{S}^*(\alpha, \beta)$ can be verified in quite a similar way as above. Finally, the case of $\mathcal{C}^*(\alpha, \beta)$ is simply obtained from that of $\mathcal{S}^*(\alpha, \beta)$ by re-replacing νa_ν instead of a_ν. □

By making use of these Lemmas, Owa [1] has derived the following results:

THEOREM 16.3. *For* $f \in \mathcal{T}$, $\mathcal{L}^\lambda f \in \mathcal{T}$ *at least for* $\lambda \geqq \lambda_*$ *with a* λ_* *less than* 2.

Proof. Since $f \in \mathcal{T}$, we have $\nu a_\nu \leqq 1$ for any $\nu \geqq 2$ by Lemma 16.3. Hence we have for

$$\mathcal{L}^\lambda f(z) = z - \sum_{\nu=2}^{\infty} \frac{a_\nu}{\nu^\lambda} z^\nu$$

the inequality

$$\sum_{\nu=2}^{\infty} \nu \frac{a_\nu}{\nu^\lambda} \leq \sum_{\nu=2}^{\infty} \frac{1}{\nu^\lambda} = \zeta(\lambda) - 1 < 1,$$

for $\lambda \geqq 2$. This shows $\mathcal{L}^\lambda f \in \mathcal{T}$ in view of Lemma 16.4. □

THEOREM 16. 4. *Let f defined by* (16.7) *be in the class $\mathcal{T}$. Then, $\mathcal{L}^\lambda f$ is in the class $\mathcal{T}^*(\alpha)$ or $\mathcal{C}(\alpha)$ for $\lambda \geqq \lambda_*$ with a λ_* less than*

$$\log \frac{2-\alpha}{1-\alpha} \Big/ \log 2 + 1 \quad or \quad \log \frac{2-\alpha}{1-\alpha} \Big/ \log 2 + 2,$$

respectively.

Proof. Since $f \in \mathcal{T}$, in view of Lemma 16.3,

$$\sum_{\nu=2}^{\infty} (\nu - \alpha) \frac{a_\nu}{\nu^\lambda} \leq \sum_{\nu=2}^{\infty} \frac{1}{\nu^\lambda} = \zeta(\alpha) - 1 < \frac{1}{2^{\lambda-2}} = 1 - \alpha$$

for any $\lambda \geqq \log((2-\alpha)/(1-\alpha))/\log 2 + 1$. This shows the part about $\mathcal{T}^*(\alpha)$. Quite similarly, the part about $\mathcal{C}$ is shown. □

THEOREM 16.5. *$f \in \mathcal{T}^*(\alpha)$ implies $\mathcal{L}^\lambda f \in \mathcal{T}^*(\alpha)$ or $\mathcal{L}^\lambda f \in \mathcal{C}(\alpha)$ at least for $\lambda \geqq \lambda_*$ with a λ_* less than* 2 *or* 3, *respectively.*

Proof. Since $f \in \mathcal{T}^*(\alpha)$, we have $(\nu - \alpha)a_\nu \leqq 1 - \alpha$ for $\nu \geqq 2$ and $0 \leqq \alpha < 1$ in view of Lemma 16.3. Hence we see that

$$\sum_{\nu=2}^{\infty} (\nu - \alpha)\frac{a_\nu}{\nu^\lambda} \leqq (1 - \alpha)\sum_{\nu=2}^{\infty} \frac{1}{\nu^\lambda} = (1 - \alpha)(\zeta(\lambda) - 1)$$

$$< \frac{1 - \alpha}{2^{\lambda-1} - 1} \leqq 1 - \alpha$$

or

$$\sum_{\nu=2}^{\infty} \nu(\nu - \alpha)\frac{a_\nu}{\nu^\lambda} \leqq \frac{2(1 - \alpha)}{2^{\lambda-1} - 1} \leqq 1 - \alpha$$

for any $\lambda \geqq 2$ or $\lambda \geqq 3$, respectively, ζ being Riemann zeta-function. This shows the results on $\mathcal{L}^\lambda f$. □

THEOREM 16.6. *Let f defined by* (16.7) *be in the class $\mathcal{T}^*(\alpha)$ or $C(\alpha)$. Then $\mathcal{L}^\lambda f$ is in the class $\mathcal{T}$ at least for $\lambda \geqq \lambda_*$ with a λ_* less than*

$$\log\frac{3 - 2\alpha}{2 - \alpha} \Big/ \log 2 + 2$$

or

$$\log\frac{3 - 2\alpha}{2 - \alpha} \Big/ \log 2 + 1,$$

respectively.

Proof. Quite similarly to Theorem 16.2. In fact, $f \in \mathcal{T}^*(\alpha)$ implies $a_\nu \leqq (1 - \alpha)/(2 - \alpha)$ for any $\nu \geqq 2$. Hence,

$$\sum_{\nu=2}^{\infty} \nu \frac{a_\nu}{\nu^\lambda} \leqq \frac{1-\alpha}{2-\alpha} \sum_{\nu=2}^{\infty} \frac{1}{\nu^{\lambda-1}}$$

$$= \frac{1-\alpha}{2-\alpha} (\zeta(\lambda - 1) - 1)$$

$$< \frac{1-\lambda}{(2-\alpha)(2^{\lambda-2} - 1)} .$$

The last member of this inequality does not exceed the unity provided $\lambda \geqq \log((3 - 2\alpha)/(2 - \alpha))/\log 2 + 2$. Hence, the result on $\mathcal{T}^*(\alpha)$ follows in view of Lemma 16.4. Next, concerning $\mathcal{C}(\alpha)$, Lemma 16.3 implies $\nu a_\nu \leqq (1 - \alpha)/(2- \alpha)$ for any $\nu \geqq 2$. Hence, we obtain

$$\sum_{\nu=2}^{\infty} \nu \frac{a_\nu}{\nu^\lambda} \leqq \frac{1-\alpha}{2-\alpha} \sum_{\nu=2}^{\infty} \frac{1}{\nu^\lambda}$$

from which the result follows similarly as above. □

With respect to the classes $\mathcal{P}^*(\alpha, \beta)$, $\mathcal{S}^*(\alpha, \beta)$, $\mathcal{C}^*(\alpha, \beta)$, the circumstances are almost similar. Accordingly, we enumerate some of the results without proof.

THEOREM 16.7. *Let f belong to $\mathcal{P}^*(\alpha, \beta)$. Then $\mathcal{L}^\lambda f$ is in the class*

$\mathcal{P}^*(\alpha, \beta)$, $\mathcal{S}^*(\alpha, \beta)$, $\mathcal{T}$, $\mathcal{T}^*(\alpha)$, or $\mathcal{C}(\alpha)$

at least for $\lambda \geqq \lambda_*$, where λ_* is a certain number less than

$$2,$$

$$2,$$

$$\log \frac{1 + 3\beta - 2\alpha\beta}{1 + \beta} \Big/ \log 2 + 1,$$

$$\log \frac{1 + 3\beta}{1 + \beta} \Big/ \log 2 + 1,$$

or

$$\log \frac{1 + 3\beta}{1 + \beta} \Big/ \log 2 + 2,$$

respectively.

THEOREM 16.8. *Let* f *defined by* (16.4) *belong to the class* $S^*(\alpha, \beta)$. *Then* $\mathcal{L}^\lambda f$ *is in the class*

$$S^*(\alpha, \beta), \quad T, \quad T^*(\alpha), \quad C(\alpha), \quad or \quad C^*(\alpha, \beta)$$

at least for $\lambda \geqq \lambda_*$ *where* λ_* *is a certain number less than*

$$2,$$

$$\log \frac{1 + 5\beta - 4\alpha\beta}{1 + 3\beta - 2\alpha\beta} \Big/ \log 2 + 2,$$

$$\log \frac{1 + 5\beta - 2\alpha\beta}{1 + 3\beta - 2\alpha\beta} \Big/ \log 2 + 2,$$

$$\log \frac{1 + 5\beta - 2\alpha\beta}{1 + 3\beta - 2\alpha\beta} \Big/ \log 2 + 3,$$

or

$$3,$$

respectively.

THEOREM 16.9. *Let f defined by* (16.4) *belong to* $C^*(\alpha, \beta)$. *Then* $\mathcal{L}^\lambda f$ *is in the class*

$$C^*(\alpha, \beta), \quad T, \quad T^*(\alpha), \quad C(\alpha), \quad P^*(\alpha, \beta), \text{ or } S^*(\alpha, \beta)$$

at least for $\lambda \geqq \lambda_*$, *where* λ_* *is a certain number less than*

$$2,$$

$$\log \frac{1 + 5\beta - 4\alpha\beta}{1 + 3\beta - 2\alpha\beta} \Big/ \log 2 + 1,$$

$$\log \frac{1 + 5\beta - 2\alpha\beta}{1 + 3\beta - 2\alpha\beta} \Big/ \log 2 + 1,$$

$$\log \frac{1 + 2\beta - \alpha\beta}{1 + 3\beta - 2\alpha\beta} \Big/ \log 2 + 2,$$

or

$$1,$$

respectively.

§ 17. Mean distortions

Let $\mathcal{F}_0$ denote the class of functions f such that $f(z)/z$ are holomorphic and non-vanishing in E. The purpose of the present section is to derive distortion relations concerning the weighted mean of some functionals defined on the general univalent class $\mathcal{S} \subset \mathcal{F}_0$ and its familiar subclasses. The functionals to be considered are $\log |tf(z)/f(zt)|$ for the classes $\mathcal{S}$ as well as $\mathcal{S}t$, and $\log |f'(z)/f'(zt)|$ for the class $\mathcal{K}$, and the mean is taken over the unit interval I with respect to t; cf. Komatu [16].

Now, let any pair of $f \in \mathcal{F}_0$ and $t \in I$ be given. Then $\log (tf(z)/f(zt))$ has, as function of z, a single-valued principal branch in E, which is determined by its value 0 at the origin. Throughout the discussions, the mean is concerned with a weighted integral of functionals which behave homogeneously with respect to f, so that the normalization is made besides $f(0) = 0$, merely by $f'(0) \neq 0$ instead of $f'(0) = 1$.

In the following lines, let p be a real-valued function absolutely integrable on I and continuous in its interior, and let $\hat{p}$ denote the mean of p over the interval $(0, t) \ominus I$ defined by

$$\hat{p}(t) = \frac{1}{t} \int_0^t p(\tau)\, d\tau.$$

We assume that $\hat{p} \geqq 0$ and $\hat{p} \not\equiv 0$ in the interior of I except in the following Lemma 17.1.

We begin with an elementary lemma:

LEMMA 17.1. *Any function* $f \in \mathcal{F}_0$ *satisfies the identity* :

$$\int_I p(t) \log \frac{t f(z)}{f(zt)} dt = \int_I \hat{p}(t) \left(\frac{zt f'(zt)}{f(zt)} - 1 \right) dt .$$

Proof. In view of $p(t) = (t\hat{p}(t))'$, the integration by parts yields

$$\int_I p(t) \log \frac{t f(z)}{f(zt)} dt$$

$$= \left[t \hat{p}(t) \log \frac{t f(z)}{f(zt)} \right]_0^1 - \int_I t \hat{p}(t) \frac{\partial}{\partial t} \log \frac{t f(z)}{f(zt)} dt$$

$$= - \int_I t \hat{p}(t) \left(\frac{1}{t} - \frac{z f'(zt)}{f(zt)} \right) dt ,$$

since the integrated part vanishes out. Hence the desired result follows. □

Based on the Lemma 17.1 just shown, we shall state several mean distortions on univalent classes.

THEOREM 17.1. *Each* $f \in \mathcal{S}$ *satisfies the mean distortion inequality*

$$\int_I p(t) \log \left| \frac{t\, f(t)}{f(zt)} \right| dt \leq \hat{p}(t) \frac{rt}{1 - rt} dt ,$$

valid for $|z| \leq r < 1$. *The equality sign appears at* z_0 *with* $|z_0| = r$ *if and only if* f *is a constant multiple of*

$$\frac{z}{((1 - \bar{z}_0 z)/|z_0|)^2} .$$

Proof. By separating the real part of the identity given in the Lemma 17.1, we get

$$\int_I p(t) \log \left| \frac{t\, f(z)}{f(zt)} \right| dt = \int_I \hat{p}(t) \left(\operatorname{Re} \frac{zt\, f'(zt)}{f(zt)} - 1 \right) dt.$$

By means of a distortion theorem of Nevanlinna [1], we have

$$\operatorname{Re} \frac{zt\, f'(zt)}{f(zt)} \leq \left| \frac{zt\, f'(zt)}{f(zt)} \right| \leq \frac{1 + rt}{1 - rt}$$

for $z \in E$ with $|z| \leq r$ and $t \in I$. In view of $\hat{p} \geq 0$, we obtain

$$\int_I \hat{p}(t) \left(\operatorname{Re} \frac{zt\, f(zt)}{f(zt)} - 1 \right) dt \leq \int \hat{p}(t) \left(\frac{1 + rt}{1 - rt} - 1 \right) dt ,$$

whence readily follows the desired inequality. The assertion on the equality sign is also transmitted from that in the Nevanlinna theorem. □

Next, let $\mathcal{S}t(\alpha)$ and $\mathcal{K}(\alpha)$ be the subclasses of $\mathcal{S}$ consisting of functions f which are starlike and convex, respectively, of order $\alpha < 1$. They are characterized by

$$\mathrm{Re}\,\frac{z f'(z)}{f(z)} > 0 \quad \text{and} \quad \mathrm{Re}\left(1 + \frac{z f''(z)}{f'(z)}\right) > \alpha$$

in E, respectively. For these subclasses, mean distortions with lower and upper estimates will be derived.

THEOREM 17.2. *Any function* $f \in St(\alpha)$ *satisfies the mean distortion inequalities*

$$-2(1-\alpha)\int_I \hat{p}(t)\frac{rt}{1+rt}\,dt \leq \int_I p(t)\log\left|\frac{t f(z)}{f(zt)}\right|dt$$

$$\leq 2(1-\alpha)\int_I \hat{p}(t)\frac{rt}{1-rt}\,dt,$$

valid for $|z| \leq r < 1$. *The left and right equality signs appear at* z_0 *with* $|z_0| = r$ *if and only if* f *is a constant multiple of* $z/(1 \pm \bar{z}_0 z/|z_0|)^2$, *respectively*.

Proof. Since for $f \in St(\alpha)$ the function $z f'(z)/f(z)$ has the real part greater than α in E and attains the value unity at the origin, the inequality of Harnack [1] yields

$$(1-\alpha)\frac{1-rt}{1+rt} \leq \mathrm{Re}\,\frac{zt\, f'(zt)}{f(zt)} - \alpha \leq (1-\alpha)\frac{1+rt}{1-rt}$$

for $|z| \leq r$ and $t \in I$. By subtracting $1-\alpha$ from each member and multiplying by $\hat{p} \geq 0$ followed by integration, we obtain the desired inequality in view of Lemma 17.1. The assertion on the equality signs is transmitted from that of the Harnack inequality. □

THEOREM 17.3. *Each function* $f \in \mathcal{K}(\alpha)$ *satisfies the mean distortion inequalities of a similar form as in Theorem* 17.2, *which is obtained by substituting* $f'(z)/f'(zt)$ *in place of* $tf(z)/f(zt)$. *The left and right equality signs appear at* z_0 *with* $|z_0| = r$ *if and only if* f *is a constant multiple of* $z/(1 \pm \bar{z}_0 z/|z_0|)$, *respectively.*

Proof. It follows from the identity

$$1 + z\frac{f''(z)}{f'(z)} = z\frac{(zf'(z))'}{zf'(z)},$$

that $f(z) \in \mathcal{K}(\alpha)$ is equivalent to $zf'(z) \in St(\alpha)$ in view of the theorem of Alexander [1]. Consequently, we see by substituting $zf'(z)$ for $f(z)$, we see that Theorem 17.1 gives the desired result. □

As an example, we consider a special function given by

$$p(t) = \rho(t;a) \equiv at^{a-1}$$

with a parameter $a > 0$, for which

$$\hat{p}(t) = t^{a-1}.$$

In this case, for instance, Theorems 17.2 and 17.3 imply the following concrete corollarly.

COROLLARY 17.1. *Let* $\Omega(z, t)$ *denote* $\log|tf(z)/f(zt)|$ *for* $f \in St(\alpha)$ *or* $\log|f'(z)/f'(zt)|$ *for* $f \in \mathcal{K}(\alpha)$. *Then the mean distortion inequalities*

$$-2(1-\alpha)\int_I \frac{rt^a}{1+rt}\,dt \leqq a\int_I t^{a-1}\Omega(z,t)\,dt$$

$$\leqq 2(1-\alpha)\int_I \frac{rt^a}{1-rt}\,dt ,$$

is valid for $|z| \leqq r < 1$, *where the assertion on equality signs is the same as in Theorem* 17.2 *or* 17.3, *respectively. The factors contained in the bounds may be expressed in the series form*

$$\mp\int_I \frac{rt^a}{1\pm rt}\,dt = \sum_{\nu=1}^{\infty} \frac{(\mp r)^\nu}{\nu + a};$$

in particular, when $a = k$ *is a positive integer, they become*

$$\sum_{\nu=1}^{\infty} \frac{(\mp r)^\nu}{\nu + k} = \frac{1}{(\mp r)^k}\left(\log\frac{1}{1\pm r} - \sum_{n=1}^{k} \frac{(\mp r)^n}{n}\right).$$

Proof. Series expansion of the integrand followed by termwise integration leads us readily to the desired result. □

Chapter 5. Distortions on real part

§ 18. Real part of analytic functions

Throughout the present chapter, we shall discuss several distortion relations concerning the real part of analytic functions f belonging to the class $\mathcal{F}$, namely, the functions f holomorphic in E and usually normalized by

$$f(0) = f'(0) - 1 = 0$$

at the origin; cf. Komatu [5, 6, 13 etc.].

By the fact that $f \in \mathcal{F}$ implies $\mathcal{L} f \in \mathcal{F}$, the discussions will be made usually in the class $\mathcal{F}$.

Again we write $f_\lambda = \mathcal{L}^\lambda f$ for the sake of brevity, and introduce two functionals $h_\lambda[f;\ r]$ and $H_\lambda[f;\ r]$ defined on $\mathcal{F}$ by

$$h_\lambda[f;\ r] := \min_{|z|=r} \operatorname{Re} \frac{f_\lambda(z)}{z},$$

$$H_\lambda[f;\ r] := \max_{|z|=r} \operatorname{Re} \frac{f_\lambda(z)}{z},$$

r being a parameter on $[0, 1)$; in particular,

$$h[f;\ r] = h_0[f;\ r] = \min_{|z|=r} \operatorname{Re}\frac{f(z)}{z},$$

$$H[f;\ r] = H_0[f;\ r] = \max_{|z|=r} \operatorname{Re}\frac{f(z)}{z}.$$

If it can be easily grasped that f is concerned, we shall abbreviate these quatities by writing

$$h_\lambda(r),\quad H_\lambda(r),\quad h(r),\quad H(r),$$

respectively.

Main problems to be discussed are to estimate h_λ or H_λ in terms of h and H. The basic tool of attack is a principle based on the theorem of Harnack [1]. Since the problem of this kind has become a clue of a series of investigations concerning the related problems performed by the present author, its discussions will be devoted in particular detail in the following lines.

Now since $f'(0) = [f(z)/z]^{z=0} = 1$, we have, in particular,

$$-\infty < h(r) \leqq 1 \leqq H(r) < \infty.$$

For the sake of brevity, we shall use an abbreviated notation:

$$\hat{f}(z) \equiv \mathcal{L} f(z) = \int_I \frac{f(zt)}{t}\, d\sigma(t)$$

and, similarly, we attach the symbol $\hat{}$ to every quantity after transformed by the operator $\mathcal{L}$; for instance, $\hat{h}$ and $\hat{H}$.

We begin with the following theorem:

THEOREM 18.1. *For any* $f \in \mathcal{F}$ *the monotonicity*

$$h(r) \leqq \hat{h}(r) \leqq 1 \leqq \hat{H}(r) \leqq H(r)$$

holds. More precisely, we have

$$\hat{h}(r) \geqq h(r) + (1 - h(r)) \int_I \frac{1 - t}{1 + t} \, d\sigma(t),$$

$$\hat{H}(r) \leqq H(r) - (H(r) - 1) \int_I \frac{1 - t}{1 + t} \, d\sigma(t)$$

for $0 \leqq r < 1$, *where the equality sign in either relation appears for an* $r \in (0, 1)$ *if and only if* $f(z) = z$, *unless* σ *is the Dirac measure at* 1.

Proof. If $f \in \mathcal{F}$, then for any fixed value of z with $|z| = r < 1$, both quantities

$$\frac{f(zt)/zt - h(r)}{1 - h(r)} \quad \text{and} \quad \frac{H(r) - f(zt)/zt}{H(r) - 1}$$

may be regarded as functions of a complex variable t holomorphic on $\{|t| \leqq 1\}$ provided $f(z) \not\equiv z$. Since they attain the value 1 at the origin and have the non-negative real part, we have in particular

$$\operatorname{Re}\frac{f(zt)/zt - h(r)}{1 - h(r)} \geqq \frac{1-t}{1+t},$$

$$\operatorname{Re}\frac{H(r) - f(zt)/zt}{H(r) - 1} \geqq \frac{1-t}{1+t}$$

for $t \in I$ and hence

$$h(rt) \geqq h(r) + (1 - h(r))\frac{1-t}{1+t},$$

$$H(rt) \leqq H(r) - (H(r) - 1)\frac{1-t}{1+t}.$$

The last relations remain to hold also when $h(r) = 1$ or $H(r) = 1$ and hence $f(z) = z$. Consequently, in view of

$$\operatorname{Re}\frac{\mathcal{L}f(z)}{z} = \int_I \operatorname{Re}\frac{f(zt)}{zt}\, d\sigma(t) \quad \text{with } \sigma(1) = 1,$$

we obtain the desired estimation. If the equality sign appears in the first estimation for an $r \in (0, 1)$, then there exists z with $|z| = r$ such that

$$\operatorname{Re}\frac{f(zt)}{zt} = h(r) + (1 - h(r))\frac{1-t}{1+t}$$

for $t \in I$. By virtue of the analyticity and the normalization at 0, the relation

$$\frac{f(zt)}{zt} = h(r) + (1 - h(r))\frac{1-t}{1+t}$$

holds for $t \in E$. Since the left-hand member is bounded

there, we must have $h(r) = 1$ and hence $f(z) = z$. The equality assertion for the second estimation is similarly verified. Or, it may be remarked that either one of the result in the Theorem follows from another by considering $2z - f(z)$ in place of $f(z)$ and replacing correspondingly $\sigma(t)$ by $2t - \sigma(t)$.

□

By the way, we remark here that the inequalities of the nature opposite to those in Theorem 18.1 can be derived:

THEOREM 18.2. *For any $f \in \mathcal{J}$*

$$\hat{h}(r) \geq H(r) - (H(r) - 1) \int_I \frac{1 + t}{1 - t} \, d\sigma(t),$$

$$\hat{H}(r) \leq h(r) + (1 - h(r)) \int_I \frac{1 + t}{1 - t} \, d\sigma(t)$$

for $r \in [0, 1)$, where the equality assertion is the same as in the preceding Theorem 18.1.

Proof. The proof proceeds similarly to that of the preceding Theorem 18.1. We only have to make use of

$$\operatorname{Re} \frac{f(zt)/zt - h(r)}{H(r) - 1} \leq \frac{1 + t}{1 - t},$$

$$\operatorname{Re} \frac{H(r) - f(zt)/zt}{1 - h(r)} \leq \frac{1 + t}{1 - t}.$$

□

Now, let $\{\mathcal{L}^{\lambda}\}_{\lambda \geq 0}$ be a family of operators possessing the structure of a semiring, that is, it satisfies the additivity condition $\mathcal{L}^{\lambda}\mathcal{L}^{\mu}=\mathcal{L}^{\lambda+\mu}$, and every $\mathcal{L}^{\lambda}$ be associated with a respective probability measure σ_{λ}. For the sake of brevity, let the image of $f\in\mathcal{F}$ by $\mathcal{L}^{\lambda}$ be denoted by f_{λ}:

$$f_{\lambda}(z):=\mathcal{L}^{\lambda}f(z)\equiv\int_{I}\frac{f(zt)}{t}\,d\sigma_{\lambda}(t).$$

The following theorems are really immediate consequences of Theorems 18.1 and 18.2.

THEOREM 18.3. *The quantities defined by*

$$h_{\lambda}(r)=\min_{|z|=r}\operatorname{Re}\frac{f_{\lambda}(z)}{z}$$

and

$$H_{\lambda}(r)=\max_{|z|=r}\operatorname{Re}\frac{f_{\lambda}(z)}{z}$$

for $r\in[0,1)$ *are increasing and decreasing in* λ, *respectively. More precisely, we have for* $\delta>0$

$$h_{\lambda+\delta}(r)\geqq h_{\lambda}(r)+\phi(\delta)(1-h_{\lambda}(r)),$$

$$H_{\lambda+\delta}(r)\leqq H_{\lambda}(r)-\phi(\delta)(H_{\lambda}(r)-1)$$

where ϕ *is given by*

$$\phi(\delta)=\int_{I}\frac{1-t}{1+t}\,d\sigma_{\delta}(t)=1-2\sum_{\nu=2}^{\infty}(-1)^{\nu}\alpha_{\nu}^{\delta},$$

the α*'s being the moments with respect to* σ. *The equality sign in either relation does not appear for any* $r \in (0, 1)$, *provided* $f(z) \not\equiv z$ *and* $\mathcal{L} f \not\equiv f$.

Proof. For any $\lambda \geqq 0$ and $\delta > 0$ we have $\mathcal{L}^{\lambda+\delta} = \mathcal{L}^{\lambda} \mathcal{L}^{\delta}$, namely

$$f_{\lambda+\delta}(z) = \int_I \frac{f_\lambda(zt)}{t} \, d\sigma_\delta(t).$$

Hence, we may take f_λ, $f_{\lambda+\delta}$ and σ_δ instead of f, $\hat{f}$ and σ in Theorem 18.1, whence readily follows the desired result. With respect to the equality sign, we have only to notice that σ_δ cannot degenerate to a point measure. □

From Theorem 18.3 just proved, we can derive other estimation with bounds of more simple nature.

THEOREM 18.4. *The quantities defined in Theorem* 17.3 *satisfy for* $\delta > 0$

$$h_{\lambda+\delta}(r) \geqq h_\lambda(r) + (1 - e^{-\phi'(0)\delta})(1 - h_\lambda(r)),$$

$$H_{\lambda+\delta}(r) \leqq H_\lambda(r) - (1 - e^{-\phi'(0)\delta})(H_\lambda(r) - 1)$$

where $\phi'(0)$ *means* $\phi'(+0)$.

Proof. Theorem 18.3 asserts that

$$\frac{h_{\lambda+\delta}(r) - h_\lambda(r)}{\delta} \geqq \frac{\phi(\delta)}{\delta}(1 - h_\lambda(r)),$$

whence follows after $\delta \to +0$ a linear differential inequality

$$\frac{\partial}{\partial \lambda} h_\lambda(r) \geqq \phi'(0)(1 - h_\lambda(r)).$$

It can be brought readily into finite form. In fact, we may write it in the form

$$\frac{\partial}{\partial \lambda}(e^{\phi'(0)\lambda} h_\lambda(r)) \geqq \phi'(0)e^{\phi'(0)\lambda}$$

which becomes, after integration with respect to λ over the interval $(\lambda, \lambda + \delta)$,

$$e^{\phi'(0)(\lambda+\delta)} h_{\lambda+\delta}(r) - e^{\phi'(0)\lambda} h_\lambda(r)$$

$$\geqq e^{\phi'(0)(\lambda+\delta)} - e^{\phi'(0)\lambda}.$$

This is the desired estimation for h. Quite similar argument applies also for H. □

Next, we have introduced in § 5 the family of operators $\{\mathcal{L}(a)^\lambda\}_{\lambda \geqq 0}$ depending on a non-negative real number a which is generated by t^a. According to the circumstances, we observe the quantities for $f \in \mathcal{F}$ defined by

$$\left.\begin{array}{l} h_\lambda(r, a) \\ H_\lambda(r, a) \end{array}\right\} = \left\{\begin{array}{l} \min \\ \max_{|z|=r} \end{array}\right. \mathrm{Re}\, \frac{\mathcal{L}(a)^\lambda f(z)}{z}.$$

The analogues of Theorems 18.3 and 18.4 can be stated as in the following forms:

THEOREM 18.5. *For any* $f \in \mathcal{J}$ *and* $\delta > 0$, *we have*

$$h_{\lambda+\delta}(r, a) \geqq h_{\lambda}(r, a) + \phi(\delta, a)(1 - h_{\lambda}(r, a)),$$

$$H_{\lambda+\delta}(r, a) \leqq H_{\lambda}(r, a) - \phi(\delta, a)(H(r, a) - 1)$$

where ϕ *is given by*

$$\phi(\delta, a) = \int_I \frac{1-t}{1+t}\, d\sigma_\delta(t, a)$$

$$= 1 - 2a^\delta \sum_{\nu=2}^{\infty} \frac{(-1)^\nu}{(\nu + a - 1)^\delta}.$$

The equality sign in either estimation does not appear for any $r \in (0, 1)$ *unless* $f(z) = z$. *If, in particular,* $a = k$ *is a positive integer,* $\phi(\delta, k)$ *is expressible in the form*

$$\phi(\delta, k) = 1 + 2(-1)^{k-1} k^\delta \left((1 - 2^{1-\delta})\, \zeta(\delta) + \sum_{\kappa=1}^{k} \frac{(-1)^{\kappa-1}}{\kappa^\delta}\right),$$

ζ *denoting Riemann zeta function.*

Proof. The inequalities having been generally shown in Theorem 18.3, it suffices to verify the expression for ϕ. We first have

$$\phi(\delta, a)$$

$$= \int_I \frac{1-t}{1+t}\, d\sigma_\delta(t; a)$$

$$= \frac{a^{\delta}}{\Gamma(\delta)} \int_I \frac{1 - t}{1 + t} t^{a-1} \left(\log \frac{1}{t}\right)^{\delta-1} dt$$

$$= \frac{a^{\delta}}{\Gamma(\delta)} \int_I \left(1 - 2 \sum_{\nu=2}^{\infty} (-1)^{\nu} t^{\nu-1}\right) t^{a-1} \left(\log \frac{1}{t}\right)^{\delta-1} dt$$

$$= 1 - 2a^{\delta} \sum_{\nu=2}^{\infty} \frac{(-1)^{\nu}}{(\nu + a - 1)^{\delta}} .$$

Next, in view of the formula

$$\sum_{\kappa=1}^{\infty} \frac{(-1)^{\kappa-1}}{\kappa^{\delta}} = (1 - 2^{1-\delta})\zeta(\delta),$$

we get for a positive integer k the relation

$$\sum_{\nu=2}^{\infty} \frac{(-1)^{\nu}}{(\nu + k - 1)^{\delta}}$$

$$= \sum_{\kappa=k+1}^{\infty} \frac{(-1)^{\kappa-k+1}}{\kappa^{\delta}}$$

$$= (-1)^{k} \left(\sum_{\kappa=1}^{\infty} - \sum_{\kappa=1}^{k} \right) \frac{(-1)^{\kappa-1}}{\kappa^{\delta}}$$

$$= (-1)^{k} (1 - 2^{1-\delta})\zeta(\delta) + (-1)^{k} \sum_{\kappa=1}^{k} \frac{(-1)^{\kappa}}{\kappa^{\delta}} .$$

By substituting this in the above expression for $\phi(\delta, a)$ with $a = k$, we obtain its desired expression. □

THEOREM 18.6. *We have for $f \in \mathcal{F}$ and $\delta > 0$*

$$h_{\lambda+\delta}(r; a) \geq h_{\lambda}(r; a) + (1 - e^{-\phi'(0,a)\delta})(1 - h_{\lambda}(r; a)),$$

$$H_{\lambda+\delta}(r; a) \leq H_{\lambda}(r; a) - (1 - e^{-\phi'(0,a)\delta})(H_{\lambda}(r; a) - 1),$$

where ϕ' is given by

$$\phi'(0, a) = \left[\frac{\partial}{\partial \delta} \phi(\delta, a)\right]^{\delta=+0}$$

$$= 2 \lim_{\delta \to +0} \sum_{\nu=2}^{\infty} (-1)^{\nu} \frac{\log(\nu + a - 1) - \log a}{(\nu + a - 1)^{\delta}}.$$

If, in particular, $a = k$ is a positive integer, then

$$\phi'(0, k) = 2(-1)^{k-1} \log \frac{k!\,(\pi/2)^{1/2}}{(2^{[k/2]}[k/2]!)^2} - \log k.$$

Proof. We first note that $\phi(+0, a) = 0$. In fact, by means of integration by parts, we get

$$\phi(\delta, a)$$

$$= \frac{a^{\delta}}{\Gamma(\delta)} \int_I \frac{1-t}{1+t} t^{a-1} \left(\log \frac{1}{t}\right)^{\delta-1} dt$$

$$= \frac{a^{\delta}}{\Gamma(\delta + 1)} \left(\left[- \frac{1 - t}{1 + t} t^{a} \left(\log \frac{1}{t} \right)^{\delta} \right]_0^1 + \int_I \frac{d}{dt} \left(\frac{1 - t}{1 + t} t^{a} \right) \left(\log \frac{1}{t} \right)^{\delta} dt \right)$$

$$= \frac{a^{\delta}}{\Gamma(\delta + 1)} \int_I \frac{d}{dt} \left(\frac{1 - t}{1 + t} t^{a} \right) \left(\log \frac{1}{t} \right)^{\delta} dt,$$

whence readily follows

$$\phi(+0, a) = \int_I \frac{d}{dt} \left(\frac{1 - t}{1 + t} t^{a} \right) dt = 0.$$

The first inequality in the Theorem yields

$$\frac{h_{\lambda+\delta}(r; a) - h_{\lambda}(r; a)}{\delta} \geqq \frac{\phi(\delta, a)}{\delta}(1 - h_{\lambda}(r; a)),$$

whence follows, as δ tends to $+0$, the inequality

$$\frac{\partial}{\partial \lambda} h_{\lambda}(r; a) \geqq \phi'(0, a)(1 - h_{\lambda}(r; a)).$$

This linear differential inequality can be brought readily into finite form. In fact, by writing it in the form

$$\frac{\partial}{\partial \lambda}(e^{\phi'(0,a)\lambda} h_{\lambda}(r; a)) \geqq \phi'(0, a) e^{\phi'(0,a)\lambda}$$

and then integrating with respect to λ over the interval $(\lambda, \lambda + \delta)$, we obtain the desired estimation for h. Similar argument applies also for H. Next, we have, in view of the expression for $\phi(\delta, a)$ given in the preceding Theorem 18.5, the relation

$$\frac{\partial}{\partial \delta} \phi(\delta, a) = 2a^{\delta} \sum_{\nu=2}^{\infty} (-1)^{\nu} \frac{\log(\nu + a - 1) - \log a}{(\nu + a - 1)^{\delta}},$$

whence follows the desired expression for $\phi'(0, a)$. Finally, if $a = k$ is a positive integer, we see that

$$\frac{\partial}{\partial \delta} \phi(\delta, k)$$

$$= 2(-1)^{k-1} k^{\delta} \log k \left((1 - 2^{1-\delta}) \zeta(\delta) - \sum_{\kappa=1}^{k} \frac{(-1)^{\kappa-1}}{\kappa^{\delta}} \right)$$

$$+ 2(-1)^{k-1} k^{\delta} \left((1 - 2^{1-\delta}) \zeta'(\delta) + 2^{1-\delta} \log 2 \cdot \zeta(\delta) \right.$$

$$\left. + \sum_{\kappa=2}^{k} \frac{(-1)^{\kappa-1}}{\kappa^{\delta}} \log \kappa \right).$$

In view of $\zeta(0) = -1/2$ and $\zeta'(0) = -(1/2)\log 2\pi$, we get

$$\phi'(0, k)$$

$$= 2(-1)^{k-1} \log k \left(\frac{1}{2} - \sum_{\kappa=1}^{k} (-1)^{\kappa-1} \right)$$

$$+ 2(-1)^{k-1} \left(\frac{1}{2} \log 2\pi - \log 2 + \sum_{\kappa=2}^{k} (-1)^{\kappa-1} \log \kappa \right),$$

which becomes the desired form, by remembering the elementary relations

$$\frac{1}{2} - \sum_{\kappa=1}^{k} (-1)^{\kappa-1} = \frac{(-1)^{k}}{2},$$

$$\sum_{\kappa=1}^{k} (-1)^{\kappa-1} \log \kappa = \log \frac{k}{(2^{[k/2]}[k/2]!)^{2}}. \qquad \square$$

§ 19. Several remarks

We have discussed in the preceding section several kinds of distortion on real part of analytic functions. In the following lines, we shall supplement some properties of the quantities

$$\phi(\delta, a) = \frac{a^{\delta}}{\Gamma(\delta)} \int_{I} \frac{1-t}{1+t}\, t^{a-1} \left(\log \frac{1}{t}\right)^{\delta-1} dt$$

and its derivative $\phi'(\delta, a)$ with respect to δ, which have appeared in Theorem 18.5.

For lower simple values of δ and a we see directly that

$$\phi(1, 1) = 2 \log 2 - 1,$$

$$\phi(2, 1) = \frac{\pi^{2}}{6} - 1,$$

$$\phi(1, 2) = 3 - 4 \log 2$$

and hence, in particular,

$$\phi(2, 1) > \phi(1, 1) > \phi(1, 2).$$

Now, we shall indicate that $\phi(\delta, a)$ shows such monotonicity in general. First, we observe the dependence of ϕ on δ:

THEOREM 19.1. *For any fixed* $a > 0$ *we have*

$$\phi(+0, a) = 0 \quad \textit{and} \quad \phi(\infty, a) = 1.$$

When δ *increases from* 0 *to* ∞, $\phi(\delta, a)$ *increases strictly from* 0 *to* 1.

Proof. The relation $\phi(+0, a) = 0$ has been shown on the way of proving Theorem 18.6. Next, we have

$$1 - \phi(\delta, a) = \frac{a^{\delta}}{\Gamma(\delta)} \int_I \frac{2t}{1+t}\, t^{a-1} \left(\log \frac{1}{t}\right)^{\delta-1} dt > 0.$$

Let any small positive number ε be given. Then, $2t/(1+t) < \varepsilon/2$ as $t < \varepsilon/4$ and hence

$$\begin{aligned} 1 - \phi(\delta, a) &< \frac{a^{\delta}}{\Gamma(\delta)} \left(\frac{\varepsilon}{2} \int_0^{\varepsilon/4} + \int_{\varepsilon/4}^1 \right) t^{a-1} \left(\log \frac{1}{t}\right)^{\delta-1} dt \\ &< \frac{a^{\delta}}{\Gamma(\delta)} \left(\frac{\varepsilon}{2} \int_I t^{a-1} \left(\log \frac{1}{t}\right)^{\delta-1} dt \right. \\ &\qquad \left. + \left(\log \frac{4}{\varepsilon}\right)^{\delta-1} \int_I t^{a-1} dt \right) \\ &= \frac{\varepsilon}{2} + \frac{1}{\Gamma(\delta)} \left(a \log \frac{4}{\varepsilon} \right)^{\delta-1}. \end{aligned}$$

In view of Stirling formula applied to $\Gamma(\delta)$, we see that

$$\frac{1}{\Gamma(\delta)}\left(a\log\frac{4}{\varepsilon}\right)^{\delta-1}$$

$$\sim\frac{1}{(2\pi\delta)^{1/2}\,e}\left(\frac{e\,a}{\delta}\log\frac{4}{\varepsilon}\right)^{\delta-1}\to 0 \quad \text{as } \delta\to\infty$$

and hence there exists $\Delta(\varepsilon)$ such that

$$1-\phi(\delta,a)<\varepsilon \quad \text{as } \delta>\Delta(\varepsilon).$$

This shows $\phi(\infty,a)=1$. Finally, let $0<\delta<\delta'$. Then

$$\phi(\delta',a)-\phi(\delta,a)$$

$$=\int_I\frac{1-t}{1+t}\,t^{a-1}\left(\frac{a^{\delta'}}{\Gamma(\delta')}\left(\log\frac{1}{t}\right)^{\delta'-1}-\frac{a^{\delta}}{\Gamma(\delta)}\left(\log\frac{1}{t}\right)^{\delta-1}\right)dt.$$

Put

$$T=\exp\left(-\frac{1}{a}\left(\frac{\Gamma(\delta')}{\Gamma(\delta)}\right)^{1/(\delta'-\delta)}\right).$$

Then, we see that as $t\ \substack{<\\>}\ T$,

$$\frac{a^{\delta'}}{\Gamma(\delta')}\left(\log\frac{1}{t}\right)^{\delta'-1}\ \substack{>\\<}\ \frac{a^{\delta}}{\Gamma(\delta)}\left(\log\frac{1}{t}\right)^{\delta-1} \quad\text{and}\quad \frac{1-t}{1+t}\ \substack{>\\<}\ \frac{1-T}{1+T}$$

and hence

$$\phi(\delta',a)-\phi(\delta,a)$$

$$> \frac{1-T}{1+T}\int_I t^{a-1}\left(\frac{a^{\delta'}}{\Gamma(\delta')}\left(\log\frac{1}{t}\right)^{\delta'-1} - \frac{a^{\delta}}{\Gamma(\delta)}\left(\log\frac{1}{t}\right)^{\delta-1}\right)dt$$

$$= 0.$$

Consequently, it has been shown that $\phi(\delta, a)$ is strictly increasing with respect to δ. □

Next, we shall observe the dependence of ϕ on a:

THEOREM 19.2. *For any fixed* $\delta > 0$ *we have*

$$\phi(\delta, +0) = 1 \quad \textit{and} \quad \phi(\delta, \infty) = 0.$$

When a *increases from* 0 *to* ∞, $\phi(\delta, a)$ *decreases strictly from* 1 *to* 0.

Proof. We see that

$$\phi(\delta, a)$$

$$= \frac{a^{\delta}}{\Gamma(\delta)}\int_I\left(t^{a-1} - \frac{2t^{a}}{1+t}\right)\left(\log\frac{1}{t}\right)^{\delta-1}dt$$

$$= 1 - \frac{2a^{\delta}}{\Gamma(\delta)}\int_I \frac{t^{a}}{1+t}\left(\log\frac{1}{t}\right)^{\delta-1}dt \to 1 \quad \text{as} \quad a \to +0,$$

since the last integral remains finite for $\delta > 0$. Or, the result could be derived more simply by means of the series form of ϕ. Now, let any small positive number ε be given. Then,

$$\frac{1-t}{1+t} < \frac{\varepsilon}{2} \quad \text{as} \quad 1 > t > \eta \equiv \frac{2-\varepsilon}{2+\varepsilon}$$

and hence

$$0 < \phi(\delta, a) < \frac{a^{\delta}}{\Gamma(\delta)}\left(\int_0^{\eta} + \frac{\varepsilon}{2}\int_{\eta}^{1}\right) t^{a-1}\left(\log\frac{1}{t}\right)^{\delta-1} dt$$

$$< \frac{a^{\delta}}{\Gamma(\delta)}\int_0^{\eta} t^{a-1}\left(\log\frac{1}{t}\right)^{\delta-1} dt + \frac{\varepsilon}{2}.$$

Since $a^{\delta} t^{a-1} \to 0$ as $a \to \infty$ uniformly for $t \in [0, \eta]$, there exists $A(\varepsilon)$ such that $0 < \phi(\delta, a) < \varepsilon$ as $a > A(\varepsilon)$. Finally, let $0 < a < a' < 1$. Then

$$\phi(\delta, a') - \phi(\delta, a)$$

$$= \frac{1}{\Gamma(\delta)}\int_I \frac{1-t}{1+t}\left(a'^{\delta} t^{a'-1} - a^{\delta} t^{a-1}\right)\left(\log\frac{1}{t}\right)^{\delta-1} dt.$$

We see that as $t \lessgtr (a/a')^{\delta/(a'-a)}$,

$$\frac{1-t}{1+t} \gtrless \frac{1-(a/a')^{\delta/(a'-a)}}{1+(a/a')^{\delta/(a'-a)}}$$

and

$$a'^{\delta} t^{a'-1} \lessgtr a^{\delta} t^{a-1}$$

and hence

$$\phi(\delta, a') - \phi(\delta, a)$$

$$< \frac{1 - (a/a')^{\delta/(a'-a)}}{1 + (a/a')^{\delta/(a'-a)}}$$

$$\frac{1}{\Gamma(\delta)} \int_I (a' t^{a'-1} - a^\delta t^{a-1}) \left(\log \frac{1}{t}\right)^{\delta-1} dt = 0.$$

Thus, it has been established that ϕ is strictly decreasing with respect to a. □

REMARK. If $\psi(t)$ is a measurable function bounded on I and left-continuous at 1, a similar argument as above for deriving $\phi(\delta, \infty) = 0$ in which $(1 - t)/(1 + t)$ is replaced by $\psi(t) - \psi(1)$ yields

$$\int_I (\psi(t) - \psi(1)) \rho_\delta(t; a) dt \to 0,$$

i. e.,

$$\int_I \psi(t) \rho_\delta(t; a) dt \to \psi(1) \quad \text{as } a \to \infty.$$

This relation corresponds to the fact that the probability density

$$\rho_\delta(t; a) = \frac{a^\delta}{\Gamma(\delta)} t^{a-1} \left(\log \frac{1}{t}\right)^{\delta-1}$$

is a kernel of singular integral tending to concentrate at $t = 1$ as $a \to \infty$.

Now, we shall denote $\phi'(0, a)$ briefly by $\Psi(a)$, namely,

$$\Psi(a) = \phi'(0, a) = \left[\frac{\partial}{\partial \delta} \phi(\delta, a) \right]^{\delta=+0}.$$

As shown in Theorem 18.6, the quantity $\Psi(k)$ with a positive integer k can be represented in terms of elementary expressions; in particular, we have

$$\Psi(1) = \log\frac{\pi}{2}, \quad \Psi(2) = \log\frac{4}{\pi}, \quad \Psi(3) = \log\frac{3\pi}{8},$$

$$\Psi(4) = \log\frac{32}{9\pi}, \quad \Psi(5) = \log\frac{45\pi}{128}, \quad \Psi(6) = \log\frac{256}{75\pi}, \quad \text{etc.}$$

We supplement here that the quantity $\Psi(a)$ shows, in general, the monotonicity with respect to a:

THEOREM 19.3. *We have*

$$\Psi(+0) = \infty \quad \textit{and} \quad \Psi(\infty) = 0.$$

When a *increases from* 0 *to* ∞, $\Psi(a)$ *decreases strictly from* ∞ *to* 0.

Proof. The expression for $\phi(\delta, a)$ obtained during the proof of Theorem 18.6 yields, after differentiation with respect to δ,

$$\frac{\partial}{\partial \delta} \phi(\delta, a)$$

$$= a^{\delta}\left(\frac{\log a}{\Gamma(\delta+1)} - \frac{\Gamma'(\delta+1)}{\Gamma(\delta+1)^2} \right) \int_I \frac{d}{dt}\left(\frac{1-t}{1+t}\, t^{a} \right) \cdot \left(\log\frac{1}{t} \right)^{\delta} dt$$

$$+ \frac{a^{\delta}}{\Gamma(\delta+1)} \int_I \frac{d}{dt}\left(\frac{1-t}{1+t}\, t^{a}\right)\left(\log\frac{1}{t}\right)^{\delta} \log\log\frac{1}{t}\, dt,$$

whence follows after integration by parts

$$\Psi(a) = \int_I \frac{d}{dt}\left(\frac{1-t}{1+t}\, t^{a}\right) \log\log\frac{1}{t}\, dt$$

$$= \int_I \frac{1-t}{1+t}\, t^{a-1} \left(\log\frac{1}{t}\right)^{-1} dt\, .$$

The decreasing property of $\Psi(a)$ is evident in view of the last expression. Now, for any $\varepsilon \in (0, 1/2)$, we get

$$\Psi(a) > \frac{1}{3}\varepsilon^{a} \int_{\varepsilon}^{1/2} t^{-1} \left(\log\frac{1}{t}\right)^{-1} dt$$

$$= \frac{1}{3}\varepsilon^{a} \left[-\log\log\frac{1}{t}\right]_{\varepsilon}^{1/2},$$

whence follows

$$\liminf_{a\to+0} \Psi(a) \geqq \frac{1}{3}\left[-\log\log\frac{1}{t}\right]_{\varepsilon}^{1/2}.$$

Since $\varepsilon \in (0, 1/2)$ is arbitrary, we conclude $\Psi(+0) = \infty$. Next, since the integrand of the above integral expressing $\Psi(a)$ is uniformly bounded on I for $a \geqq 2$ and tends to 0 as $a \to \infty$, it follows that $\Psi(\infty) = 0$. □

§ 20. Bounds on real part

We have derived in § 18 two kinds of distortion inequalities for the quantities

$$h_\lambda(r) = \min_{|z|=r} \operatorname{Re} \frac{f_\lambda(z)}{z}$$

and

$$H_\lambda(r) = \max_{|z|=r} \operatorname{Re} \frac{f_\lambda(z)}{z}$$

in Theorems 18.3 and 18.4. The former asserts that

$$h_{\lambda+\delta}(r) \geqq h_\lambda(r) + \phi(\delta)(1 - h_\lambda(r)),$$

$$H_{\lambda+\delta}(r) \leqq H_\lambda(r) - \phi(\delta)(H_\lambda(r) - 1),$$

while the latter asserts that

$$h_{\lambda+\delta}(r) \geqq h_\lambda(r) + (1 - e^{-\phi'(0)\delta})(1 - h_\lambda(r)),$$

$$H_{\lambda+\delta}(r) \leqq H_\lambda(r) - (1 - e^{-\phi'(0)\delta})(H_\lambda(r) - 1).$$

for $\delta > 0$.

As seen from the proof of Theorem 19.4, the latter inequalities are obtained from the former by a mere limitting process. So, the latter may be regarded as weaker than the former.

There are two possibilities:

$$\text{(A)} \qquad \phi(\delta) \geqq 1 - e^{-\phi'(0)\delta}$$

or

$$\text{(B)} \qquad \text{otherwise.}$$

Based on the above-mentioned reason, it seems to be natural to expect that the case (A) will appear, provided the class

$$\{h_\lambda(r)\} \quad (\delta > 0,\ r \in I,\ f \in \mathcal{F})$$

is wide enough.

But, in general, (A) is invalid without any restriction: for instance, the quantity

$$\phi_0(\delta) = 1 - e^{\alpha\delta} - \delta^2$$

satisfies

$$1 - e^{-\phi_0'(0)\delta} = 1 - e^{\alpha\delta}$$

$$> 1 - e^{\alpha\delta} - \delta^2 = \phi_0(\delta).$$

Accordingly, a question arises for giving a condition necessary and/or sufficient in terms of the class in order that the case (A) really holds. If (B) holds, the estimates given in Theorem 18.3 would be ameriolated at least partly by those in Theorem 18.4.

By the way, we have remarked that inequalities of the nature opposite to those in Theorem 19.1 can be derived. Corresponding to those in Theorem 19.3, it is mentioned also in the following form:

THEOREM 20.1. *The quantities appearing in Theorem* 17.3 *satisfy the estimations*

$$h_{\lambda+\delta}(r) \geqq H_{\lambda}(r) - (H_{\lambda}(r) - 1) \int_{I} \frac{1+t}{1-t}\, d\sigma_{\delta}(t),$$

$$H_{\lambda+\delta}(r) \leqq h_{\lambda}(r) + (1 - h_{\lambda}(r)) \int_{I} \frac{1+t}{1-t}\, d\sigma_{\delta}(t)$$

for $\delta > 0$ *and* $0 \leqq r < 1$, *where the equality sign in either relation does not appear for any* $r \in (0, 1)$ *provided* $f(z) \not\equiv z$ *and* $\mathcal{L} f \not\equiv f$.

Proof. The essential part has been proved in Theorem 18.2. In fact, we have only to take f_{λ}, $f_{\lambda+\delta}$ and σ_{δ} instead of f, f' and σ, respectively, in Theorem 17.2. With respect to the equality sign, the circumstances are the same as in Theorem 18.3. □

The analogue of Theorem 20.1 can be obtained corresponding to Theorem 18.5 also in the following form:

THEOREM 20.2. *For any* $f \in \mathcal{F}$ *and* $\delta > 0$, *we have*

$$h_{\lambda+\delta}(r, a) \geqq H_{\lambda}(r, a) - (H_{\lambda}(r, a) - 1) \int_{I} \frac{1+t}{1-t}\, d\sigma_{\delta}(t),$$

$$H_{\lambda+\delta}(r, a) \leqq h_\lambda(r, a) + (1 - h_\lambda(r, a)) \int_I \frac{1+t}{1-t}\, d\sigma_\delta(t),$$

where the factor in the right-hand sides associated with the integral may be also written as

$$\int_I \frac{1+t}{1-t}\, d\sigma_\delta(t, a) = 1 + 2a^\delta \sum_{\nu=2}^{\infty} \frac{1}{(\nu + a - 1)^\delta}.$$

The equality sign in either estimation does not appear for any $r \in (0, 1)$ *unless* $f(z) = z$. *If, in particular,* $a = k$ *is a positive integer, the above integral factor is also expressible in the form*

$$\int_I \frac{1+t}{1-t}\, d\sigma_\delta(t, a) = 1 + 2k^\delta \left(\zeta(\delta) - \sum_{\kappa=1}^{k} \frac{1}{\kappa^\delta}\right),$$

ζ *denoting Riemann zeta function.*

Proof. The inequalities have been established in Theorem 19. 1. With respect to the integral factor, we have

$$\int_I \frac{1+t}{1-t}\, d\sigma_\delta(t, a)$$

$$= \frac{a^\delta}{\Gamma(\delta)} \int_I \frac{1+t}{1-t}\, t^{a-1} \left(\log \frac{1}{t}\right)^{\delta-1} dt$$

$$= \frac{a^\delta}{\Gamma(\delta)} \int_I \left(1 + 2 \sum_{\nu=2}^{\infty} t^{\nu-1}\right) t^{a-1} \left(\log \frac{1}{t}\right)^{\delta-1} dt$$

$$= 1 + 2a^{\delta} \sum_{\nu=2}^{\infty} \frac{1}{(\nu + a - 1)^{\delta}} .$$

Next, we get for a positive integer k the relation

$$\sum_{\nu=2}^{\infty} \frac{1}{(\nu + k - 1)^{\delta}} = \left(\sum_{\kappa=1}^{\infty} - \sum_{\kappa=1}^{k} \right) \frac{1}{\kappa^{\delta}}$$

$$= \zeta(\delta) - \sum_{\kappa=1}^{k} \frac{1}{\kappa^{\delta}} .$$

By substituting this for the above expression, we obtain the desired result. □

§ 21. Distortions by a differential operator

Let $\tilde{\mathcal{F}}$ denote the class of analytic functions f which are holomorphic in the unit disk $E = \{|z| < 1\}$. Let $\mathcal{F}$ and $\mathcal{G}$ be its subclasses consisting of $f \in \tilde{\mathcal{F}}$ normalized by $f(0) = f'(0) - 1 = 0$ and $f(0) = 1$ at the origin, respectively.

We shall discuss in the present section a differential operator defined by

$$\Lambda \equiv \Lambda_{\alpha} = \Theta(\alpha) := 1 + \alpha \frac{d}{d \log z}$$

$$= 1 + \alpha z \frac{d}{dz}$$

on $\tilde{\mathcal{F}}$, α being a real positive parameter; cf. Komatu [17].

Miller [1] has shown that for $f \in \tilde{\mathcal{F}}$, Re $\Lambda f(z) > 0$ implies Re $f(z) > 0$, and Altintas [1] has given a proof of the equivalent fact in the form that Re $\Lambda f(z) > \beta$ implies Re $f(z) > \beta$, where β is any real constant satisfying

$$\beta < \text{Re } f(0).$$

The main purpose of the present section is to sharpen this result into precise form together with exact extremality assertion. As a direct consequence of our main Theorem, we derive a precise estimation on Re $f(z)^{\eta}$ with any $\eta \in (0, 1)$ implied by Re $\Lambda f(z) > \beta$ which involves the improvement of a result given by Owa and Zhwo-Ren Wu [1].

We now begin with an elementary Lemma which may really be regarded as a particular case of a theorem given by Chichra [1]:

LEMMA 21.1. *For any $f \in \mathcal{F}$ we have the identity*

$$f(z) = \frac{1}{\alpha} z^{-1/\alpha} \int_0^z \zeta^{1/\alpha-1} \Lambda f(\zeta)\, d\zeta$$

$$= \frac{1}{\alpha} \int_0^1 t^{1/\alpha-1} \Lambda f(zt)\, dt,$$

where the power functions denote the principal branches and the integration with respect to ζ *is taken along the segment from* 0 *to* z .

Proof. From the relation

$$\Lambda f(z) = \alpha z^{1-1/\alpha} \frac{d}{dz} z^{1/\alpha} f(z)$$

we obtain

$$f(z) = \frac{1}{\alpha} z^{-1/\alpha} \left(\int_0^z \zeta^{1/\alpha-1} \Lambda f(\zeta)\, d\zeta + C \right).$$

In view of the regularity of f at the origin, the integration constant C must vanish, whence follows the first expression. The change of variable by means of $\zeta = zt$ yields the second expression. □

Our main result involving the differential operator Λ is contained in the following theorem:

THEOREM 21.1. *For any* $f \in \mathcal{F}$, *write* $f(0) = a_0 + ib_0$, *and choose* $\beta < a_0$. *If* $\mathrm{Re}\, \Lambda f(z) > \beta$, *then*

$$\mathrm{Re}\, f(z) > \beta + (a_0 - \beta)\, q(\alpha),$$

where

$$q(\alpha) := \frac{1}{\alpha} \int_0^1 t^{1/\alpha-1} \frac{1-t}{1+t}\, dt > 0.$$

The function $\hat{f}$ *of the form*

$$\hat{f}(z) = \beta + \frac{\hat{a_0} - \beta}{\alpha} \int_0^1 t^{1/\alpha - 1} \frac{1 - \varepsilon zt}{1 + \varepsilon zt} \, dt + i\hat{b_0}$$

with $\hat{f}(0) = \hat{a_0} + i\hat{b_0}$ *and* $|\varepsilon| = 1$ *is extremal in the following sense* :

$$\inf_{z \in E} \operatorname{Re} \Lambda \hat{f}(z) = \beta,$$

$$\inf_{z \in E} \operatorname{Re} \hat{f}(z) = \beta + (\hat{a_0} - \beta) q(\alpha).$$

Proof. Since $\Lambda f(z) = f(z) + \alpha z f'(z)$ satisfies the conditions $\operatorname{Re} \Lambda f(z) > \beta$ in E and $\operatorname{Re} \Lambda f(0) = \operatorname{Re} f(0) = a_0$, the Harnack inequality yields

$$\operatorname{Re} \frac{\Lambda f(\zeta) - \beta}{a_0 - \beta} \geq \frac{1 - |\zeta|}{1 + |\zeta|}$$

for $\zeta \in E$, whence follows the inequality

$$\operatorname{Re} \Lambda f(zt) > \beta + (a_0 - \beta) \frac{1 - t}{1 + t}$$

for $z \in E$ and $t \in (0, 1]$. Consequently, in view of Lemma 21.1, we obtain

$$\operatorname{Re} f(z) = \frac{1}{\alpha} \int_0^1 t^{1/\alpha - 1} \operatorname{Re} \Lambda f(zt) \, dt$$

$$> \frac{1}{\alpha} \int_0^1 t^{1/\alpha - 1} \left(\beta + (a_0 - \beta) \frac{1 - t}{1 + t} \right) dt$$

$$= \beta + (a_0 - \beta) q(\alpha).$$

Next, by substituting $\zeta = zt$, we have

$$\hat{f}(z)$$

$$= \beta + \frac{\hat{a_0} - \beta}{\alpha} z^{-1/\alpha} \int_0^z \zeta^{1/\alpha - 1} \frac{1 - \varepsilon\zeta}{1 + \varepsilon\zeta} d\zeta + ib_0$$

and

$$\Lambda \hat{f}(z)$$

$$= \alpha z^{1-1/\alpha} \left(\frac{\beta + i\hat{b_0}}{\alpha} z^{1/\alpha - 1} + \frac{\hat{a_0} - \beta}{\alpha} z^{1/\alpha - 1} \frac{1 - \varepsilon z}{1 + \varepsilon z} \right)$$

$$= \beta + (\hat{a_0} - \beta) \frac{1 - \varepsilon z}{1 + \varepsilon z} + i\hat{b_0}.$$

The infima of the real part of these functions are both attained at the boundary point $z = \bar{\varepsilon}$, where they behave holomorphically after analytic continuation. □

As a direct consequence of Theorem 21.1, we mention here the following theorem:

THEOREM 21.2. *Under the notation of the preceding Theorem* 21.1, *assume that* $f \in \mathcal{F}$ *satisfies*

$$\beta + (a_0 - \beta) q(\alpha) \geq 0.$$

If $\operatorname{Re} \Lambda f(z) > \beta$, *then,*

$$\operatorname{Re} f(z)^{\eta} > (\beta + (a_0 - \beta)q(\alpha))^{\eta} \quad \textit{for any} \ \eta \in (0, 1),$$

where the branch of $f(z)^{\eta}$ *is taken in such a way that the absolute value of argument is the smallest at the origin. The extremal function is given by* $\hat{f}$ *in Theorem* 21.1 *with* $\hat{b_0} = 0$.

Proof. Since in view of Theorem 21.1

$$\operatorname{Re} f(z) > \beta + (a_0 - \beta)q(\alpha) \geqq 0,$$

f vanishes nowhere in E and hence the designated branch of $f(z)^{\eta}$ is surely determined as a single-valued function. If put $f(z) = Re^{i\theta}$, then

$$\theta \in \left(-\frac{\pi}{2}, \frac{\pi}{2}\right) \quad \text{and} \quad \operatorname{Re} f(z)^{\eta} = R^{\eta} \cos \eta\theta.$$

Now, for $\eta \in (0, 1)$, since

$$\varphi(\theta) = \cos \eta\theta - \cos^{\eta}\theta$$

satisfies $\varphi(0) = 0$ and

$$\varphi'(\theta) = \eta\left(\frac{\sin\theta}{\cos^{1-\eta}\theta} - \sin\eta\theta\right) > 0$$

in the interval $(0, \pi/2)$, it increases strictly with respect to $\theta \in [0, \pi)$. Consequently, we see that $\cos\eta\theta \geq \cos^{\eta}\theta$ for $\theta \in (-\pi/2, \pi/2)$ and

$$\operatorname{Re} f(z)^{\eta} = R^{\eta}\cos\eta\theta \geqq (R\cos\theta)^{\eta} = (\operatorname{Re} f(z))^{\eta},$$

where the equality sign in the intermediate inequality holds merely for $\theta = 0$. The function $\hat{f}$ given in Theorem 20.1 attains the value $\beta + (\hat{a}_0 - \beta) q(\alpha)$ on the real axis at a boundary point of E only if $\hat{b}_0 = 0$. Thus we obtain the desired result including the extremality assertion. □

Now, for the result of Miller and Altintas, Chichra [1] has noticed that, more generally, the same statement remains valid for complex values of α also provided Re $\alpha > 0$. Corresponding to Theorem 21.1, the Chichra's result can be also improved as follows; cf. Komatu [21]:

THEOREM 21.3. *For any* $f \in \mathcal{F}$, *write* $f(0) = a_0 + ib_0$ *and choose* $\beta < a_0$. *If* $\operatorname{Re}(f'(z) + \alpha z f''(z)) > \beta$, *then*

$$\operatorname{Re} f'(z) > \beta + (1 - \beta) q(u) \quad \textit{with} \quad u = \operatorname{Re} \alpha,$$

where q *is given by*

$$q(0) = 0, \quad q(u) = \frac{1}{u} \int_I t^{1/u} \frac{1 - t}{1 + t} dt > 0 \quad (u > 0).$$

Proof. Based on the identity

$$f'(z) + \alpha z f''(z) = \alpha z^{1-1/\alpha} (z^{1/\alpha} f'(z))'$$

we get the relation

$$f'(z) = \frac{1}{\alpha} z^{-1/\alpha} \int_0^z \zeta^{1/\alpha - 1} (f'(\zeta) + \alpha\zeta f''(\zeta)) d\zeta$$

$$= \frac{1}{\alpha} \int_I t^{1/\alpha} (f'(zt) + \alpha z t f''(zt)) \, dt ,$$

where the power functions denote the principal branches and the integration with respect to ζ is taken along the segment. First, suppose that α is imaginary and Re $\alpha > 0$. and put $t = \tau^{\alpha}$. In view of Re $\alpha > 0$ we have Re $\alpha^{-1} > 0$ and hence $|\tau| = t^{\mathrm{Re}\,\alpha^{-1}}$ yields that $t \in I$ implies $|\tau| < 1$. By putting $\alpha = u + iv$, we have

$$\log |\tau| = \frac{u}{u^2 + v^2} \log t , \quad \arg \tau = \frac{-v}{u^2 + v^2} \log t .$$

When t moves from 0 to 1 along the segment I, then τ moves from 0 to 1 along an arc S of the logarithmic spiral $u \arg \tau + v \log |\tau| = 0$ which lies on the closed unit disk on the τ-plane. By means of this change of variable, we obtain

$$f'(z) = \int_S (f'(z\tau^{\alpha}) + \alpha z \tau^{\alpha} f''(z\tau^{\alpha})) \, d\tau.$$

As $t \to +0$, τ moves on the spiral toward 0 winding around the origin in positive or negative sense accrrding to $v > 0$ or $v < 0$, respectively. Though the integrand, qua function of τ, is multi-valued because of the logarithmic branch point lying at $\tau = 0$, it converges uniformly to a definite value $f'(0) = 1$ as τ near terminal part of S tends to 0, and the length of S is finite. Consequently, we may replace the integration path S by the segment I on the τ-plane, whence follows

$$\mathrm{Re}\ f'(z) = \int_I \mathrm{Re}\ (f'(z\tau^\alpha) + \alpha z\tau^\alpha f''(z\tau^\alpha))\,d\tau.$$

If α is real, the last relation is evident since S then degenerates to I. Now, it follows from the Harnack inequality that the assumption yields

$$\mathrm{Re}\ (f'(\zeta) + \alpha\zeta f''(\zeta) > \beta + (1 - \beta)\frac{1 - |\zeta|}{1 + |\zeta|}$$

for $|\zeta| < 1$. Therefore, we obtain the desired inequality with

$$q\,(u) = \int_I \frac{1 - \tau^u}{1 + \tau^u}\,d\tau = \int_I t^{1/u}\,\frac{1 - t}{1 + t}\,dt\,.$$

Finally, in case $u = 0$, we have only to consider the limit process. In fact, the statement for $u = \mathrm{Re}\ \alpha > 0$ then reduces to the desired one after $u \to +0$. □

Now, we note here that the factor $\alpha^{-1} t^{1/\alpha - 1}$ involved in the integrand of the expression for $q\,(\alpha)$ which appeared in the Theorem 21.1 is a probability density on the unit interval I which has been denoted previously by $\rho(t\,;\,\alpha^{-1})$ and dealt with in detail in § 5.

For instance, since integration by parts yields

$$q\,(\alpha) = 2\int_I \frac{t^{1/\alpha}}{(1 + t)^2}\,dt\,,$$

integrand of the expression for $q(\alpha)$ is a probability density on the unit interval I which has been denoted previously by $\rho(t;\alpha^{-1})$ and dealt with in detail in § 5. For instance, since the integration by parts yields

$$q(\alpha) = 2\int_I \frac{t^{1/\alpha}}{(1+t)^2}\,dt,$$

we see that $q(\alpha)$ is increasing in $(0,\infty)$, $q(+0) = 0$ and and $q(\infty) = 1$.

On the other hand, if α is, in particular, the reciprocal of a positive integer k, then q becomes

$$q\left(\frac{1}{k}\right) = k\int_I t^{k-1}\,\frac{1-t}{1+t}\,dt$$

and can be elementarily calculated. For lower values of k, we have

$$q(1) = 2\log 2 - 1, \qquad q\left(\frac{1}{2}\right) = 3 - 4\log 2,$$

$$q\left(\frac{1}{3}\right) = 6\log 2 - 4, \qquad q\left(\frac{1}{4}\right) = \frac{17}{3} - 8\log 2.$$

We shall here add two remarks; cf. Komatu [18].

First, S. Owa and C. Y. Shen have reported a related result at the Annual Meeting of the Math. Soc. of Japan held March/ April, 1988. It may be mentioned that if $f \in \mathcal{G}$, then

$$\mathrm{Re}\,(f(z) + \alpha f'(z)) > \beta$$

implies

$$\operatorname{Re} f(z) > \frac{\alpha + 2\beta}{\alpha + 2} = \beta + (1 - \beta)\frac{\alpha}{\alpha + 2}.$$

Though the lower bound of $\operatorname{Re} f(z)$ in this estimation gives a fairly good approximation, it is (as a matter of course) smaller than the exact value $\beta + (1 - \beta) q(\alpha)$.

This can be directly verified. In fact, integration by parts yields

$$\frac{\alpha + 2}{\alpha} q(\alpha)$$

$$= \frac{2}{\alpha}\int_I \left(\frac{1}{\alpha} + \frac{1}{2}\right) t^{1/\alpha - 1/2} \frac{t^{-1/2}(1 - t)}{1 + t}\, dt$$

$$= \frac{2}{\alpha}\left(-\int_I t^{1/\alpha + 1/2} \frac{d}{dt}\frac{t^{-1/2}(1 - t)}{1 + t}\, dt\right)$$

$$= \frac{1}{\alpha}\int_I t^{1/\alpha - 1}\, dt + \frac{2}{\alpha}\int_I t^{1/\alpha}\frac{1 - t}{(1 + t)^2}\, dt.$$

The first term of the last member is equal to unity while its second term is positive, whence follows the desired inequality

$$\frac{\alpha + 2}{\alpha} q(\alpha) > 1,$$

i. e., the inequality

$$\frac{\alpha + 2\beta}{\alpha + 2} < \beta + (1 - \beta)\, q(\alpha)$$

holds for any $\beta \in (-\infty, 1)$ and $\alpha \in (0, \infty)$.

Second, Owa and Zhwo-Ren Wu [1] have given in connection with Theorem 20.2 a lower bound of Re $f(z)^{1/2}$. It is also readily verified that this bound is smaller than the exact value. The inequality to be verified is equivalent to

$$q(\alpha) > \frac{1}{1 - \beta}\left(\left(\frac{\alpha + (\alpha^2 + 4\beta(1 + \alpha))^{1/2}}{2(1 + \alpha)}\right)^2 - \beta\right)$$

$$= \frac{2\alpha}{1 + \alpha}\,\frac{\beta}{(\alpha^2 + 4\beta(1 + \alpha))^{1/2} - \alpha + 2\beta(1 + \alpha)}\,.$$

For the sake of brevity, denoting by $R(\alpha, \beta)$ the last member of this expression, we get

$$R(\alpha, 1) = \frac{\alpha}{(1 + \alpha)(2 + \alpha)} < \frac{\alpha}{\alpha + 2}$$

and

$$\frac{\partial}{\partial \beta} R(\alpha, \beta)$$

$$= \frac{2\alpha}{1 + \alpha}$$

$$\frac{\alpha^2 + 2\beta(1 + \alpha) - \alpha(\alpha^2 + 4\beta(1 + \beta))^{1/2}}{((\alpha^2 + 4\beta(1 + \alpha))^{1/2} - \alpha + 2\beta(1 + \alpha))^2\,(\alpha^2 + 4\beta(1 + \alpha))^{1/2}}$$

$$> 0$$

for any $\alpha \in (0, \infty)$. Hence, we really have

$$R(\alpha, \beta) < \frac{\alpha}{\alpha + 2} < q(\alpha)$$

for any $\beta \in [0, 1)$ and $\alpha \in (0, \infty)$.

In conclusion, we illustrate our result by assigning numerical values of the bounds for a few pairs of (α, β): $(1, 1/2)$, $(1, 1/4)$ and $(1/2, 1/2)$.

The lower bound of $\mathrm{Re}\, f(z)^{1/2}$ derived by Owa and Wu gives

$$\frac{1 + 5^{1/2}}{4} \doteqdot 0.8090, \qquad \frac{1 + 3^{1/2}}{4} \doteqdot 0.6830,$$

$$\frac{1 + 13^{1/2}}{6} \doteqdot 0.7676,$$

respectively, while the values of exact lower bound are

$$(\log 2)^{1/2} \doteqdot 0.8325, \qquad \left(\frac{3 \log 2 - 1}{2}\right)^{1/2} \doteqdot 0.7346,$$

$$(2(1 - \log 2))^{1/2} \doteqdot 0.7835,$$

respectively. Similarly, the lower bound of $\mathrm{Re}\, f(z)$ derived by Owa and Shen gives

$$\frac{2}{3} \doteqdot 0.6667, \quad \frac{1}{2} = 0.5, \quad \frac{3}{5} = 0.6,$$

respectively, while the values of exact lower bound are

$$\log 2 \doteqdot 0.6931, \qquad \frac{3\log 2 - 1}{2} \doteqdot 0.5397,$$

$$2(1 - \log 2) \doteqdot 0.6138,$$

respectively.

§ 22. Generalizations to higher order

We have introduced in § 5 and observed often in subsequent sections an integral operator $\mathcal{L}(a)$ defined on $\mathcal{F}$ which is represented by

$$\mathcal{L}(a)f(z) = a\int_I t^{a-2} f(zt)\,dt$$

where a is a positive parameter.

On the other hand, as seen in the preceding section, Miller and Altintas have discussed a differential operator defined on $\mathcal{G}$ by

$$\Lambda(\alpha) = 1 + \alpha\frac{d}{d\log z}$$

where α is a positive parameter.

As indicated already, the defining representation of $\mathcal{L}(a)$ shows that the normalization $f'(0) = 1$ is inessential and further it is applicable to any function of $\tilde{\mathcal{F}}$ provided $a > 1$. We shall deal with the differential operator in intimate relation to $\mathcal{L}(a)$ from a general standpoint; for the following discussions, cf. Komatu [21].

The interrelation between Λ and $\mathcal{L}$, which generalizes Theorem 5.1, is given by the following Theorem and its Corollary:

THEOREM 22.1. *The differential operator* $\Lambda(\alpha)$ *is the inverse of the integral operator* $(\alpha+1)^{-1}\mathcal{L}((\alpha+1)/\alpha)$, *namely,*

$$\frac{1}{\alpha+1}\mathcal{L}\left(\frac{\alpha+1}{\alpha}\right)\Lambda(\alpha) = \mathrm{id},$$

or, in other words, $(a(a-1)^{-1})\Lambda((a-1)^{-1})$ *is the inverse of* $\mathcal{L}(a)$.

Proof. Direct calculation yields

$$\Lambda(\alpha)\frac{1}{\alpha+1}\mathcal{L}\left(\frac{\alpha+1}{\alpha}\right)f(z)$$

$$= \Lambda(\alpha)\frac{1}{\alpha}\int_I t^{1/\alpha-1} f(zt)\,dt$$

$$= \frac{1}{\alpha}\int_I t^{1/\alpha-1}(f(zt)+\alpha zt\, f'(zt))\,dt$$

$$= \int_I \frac{\partial}{\partial t}(t^{1/\alpha} f(zt))\,dt = f(z).$$ □

Theorem 22.1 leads us to the following Corollary which may be regarded as a relation defining the fractional integral operator $\Lambda(\alpha)^{-\lambda}$.

COROLLARY 22.1. *If* $\alpha > 0$ *and* $\lambda \geqq 0$, *we have*

$$\frac{1}{(\alpha + 1)^\lambda} \mathcal{L}\left(\frac{\alpha + 1}{\alpha}\right)^\lambda \Lambda(\alpha)^\lambda = \mathrm{id}.$$

We noticed in Theorem 11.1 the analytic prolongability of $\mathcal{L}(a)^\lambda$. In view of Corollary 21.1, the operator $\Lambda(\alpha)^\lambda$ also possesses a similar prolongability with respect to the pair of parameters λ and a.

In the following lines we restrict ourselves to the case $\lambda > 0$ and $\alpha > 0$. While in the preceding section we have partly improved the result obtained by Miller and Altintas that if $f \in \mathcal{G}$ satifies $\mathrm{Re}\, \Lambda(\alpha) f(z) > \beta$, then $\mathrm{Re}\, f(z) > \beta$, it can be further generalized as in the following form:

THEOREM 22.2. *If* $\tilde{f} \in \mathcal{J}$ *with* $f(0) = a_0 + ib_0$ *and* $a_0 > \beta_*$ *satisfies in* E *the inequality* $\mathrm{Re}\, \Lambda(\alpha)^\lambda f(z) > \beta_*$, *then*

$$\mathrm{Re}\, f(z) > \beta_* + (a_0 - \beta_*)\Psi(\lambda, \alpha),$$

where

$$\Psi(\lambda, \alpha) = \frac{1}{\alpha^\lambda \Gamma(\lambda)} \int_I t^{1/\alpha - 1} \left(\log \frac{1}{t}\right)^{\lambda - 1} \frac{1 - t}{1 + t}\, dt .$$

The function $\hat{f}$ *of the form*

$\hat{f}(z;\beta)$

$$= \beta + \frac{\hat{a_0} - \beta}{\alpha^{\lambda}\Gamma(\lambda)} \int_I t^{1/\alpha - 1}\left(\log\frac{1}{t}\right)^{\lambda-1} \frac{1 - \varepsilon z t}{1 + \varepsilon z t}\, dt + i\,\hat{b_0}$$

with $\hat{f}(0;\beta) = \hat{a_0} + i\,\hat{b_0}$ *and* $|\varepsilon| = 1$ *is extremal in the following sense* :

$$\inf_{z\in E} \operatorname{Re} \Lambda(\alpha)^{\lambda}\hat{f}(z;\beta_*) = \beta_*,$$

$$\inf_{z\in E} \operatorname{Re} \hat{f}(z;\beta_*) = \beta_* + (\hat{a_0} - \beta_*)\,\Psi(\lambda,\alpha).$$

Proof. We first note that

$$\mathcal{L}(a)^{\lambda}\,1 = \frac{a^{\lambda}}{\Gamma(\lambda)} \int_I t^{a-2}\left(\log\frac{1}{t}\right)^{\lambda-1} dt = \left(\frac{a}{a-1}\right)^{\lambda}$$

and hence, in view of analytic prolongability, for any constant c,

$$\Lambda(\alpha)^{\lambda} c = \frac{1}{(\alpha+1)^{\lambda}}\,\mathcal{L}\left(\frac{\alpha+1}{\alpha}\right)^{\lambda} c$$

$$= \frac{1}{(\alpha+1)^{\lambda}}(\alpha+1)^{\lambda} c = c\,.$$

Consequently, Harnack inequality yields

$$\operatorname{Re}\frac{\Lambda(\alpha)^{\lambda} f(\zeta) - \beta_*}{a_0 - \beta_*} \geq \frac{1 - |\zeta|}{1 + |\zeta|}$$

for $\zeta \in E$, whence follows

$$\operatorname{Re} \Lambda(\alpha)^{\lambda} f(zt) > \beta_{*} + (a_0 - \beta_{*}) \frac{1 - t}{1 + t}$$

for $z \in E$ and $t \in (0, 1]$. Consequently, based on the relation

$$\begin{aligned} f(z) &= \Lambda(\alpha)^{-\lambda} \Lambda(\alpha)^{\lambda} f(z) \\ &= \frac{1}{(\alpha + 1)^{\lambda}} \mathcal{L}\left(\frac{\alpha + 1}{\alpha}\right)^{\lambda} \Lambda(\alpha)^{\lambda} f(z) \\ &= \frac{1}{\alpha^{\lambda}\Gamma(\lambda)} \int_I t^{1/\alpha - 1}\left(\log \frac{1}{t}\right)^{\lambda - 1} \Lambda(\alpha)^{\lambda} f(zt)\, dt, \end{aligned}$$

we obtain

$$\begin{aligned} &\operatorname{Re} f(z) \\ &> \frac{1}{\alpha^{\lambda}\Gamma(\lambda)} \int_I t^{1/\alpha - 1}\left(\log \frac{1}{t}\right)^{\lambda - 1} (\beta_{*} + (a_0 - \beta_{*})) \frac{1 - t}{1 + t}\, dt \\ &= \beta_{*} + (a_0 - \beta_{*}) \Psi(\lambda, \alpha). \end{aligned}$$

Next, the function $\hat{f}$ given in the theorem may be written in the form

$$\hat{f}(z; \beta) = \beta + (\hat{a_0} - \beta) \Lambda(\alpha)^{-\lambda} \frac{1 - \varepsilon z}{1 + \varepsilon z} + i \hat{b_0},$$

whence follows

$$\Lambda(\alpha)^{\lambda} \hat{f}(z;\beta) = \beta + (\hat{a_0} - \beta)\frac{1-\varepsilon z}{1+\varepsilon z} + i\,\hat{b_0}.$$

Thus, the last assertion follows readily. □

COROLLARY 22.2. *If* $f \in \mathcal{F}$ *with* $f(0) = a_0 + ib_0$ *and* $a_0 < \beta^*$ *satisfies in* E *the inequality* $\text{Re}\, \Lambda(\alpha)^{\lambda} f(z) < \beta^*$, *then*

$$\text{Re}\, f(z) < \beta^* - (\beta^* - a_0)\, \Psi(\lambda, \alpha),$$

where Ψ *is the expression defined in Theorem* 22.2. *The function* f *given in Theorem* 22.2 *is extremal in the sense:*

$$\sup_{z \in E} \text{Re}\, \Lambda(\alpha)^{\lambda} \hat{f}(z;\beta^*) = \beta^*,$$

$$\sup_{z \in E} \text{Re}\, \hat{f}(z;\beta^*) = \beta^* - (\beta^* - \hat{a}_0)\, \Psi(\lambda, \alpha).$$

Proof. We have only to apply Theorem 21.2 to the function $-f$ instead of f and to replace the quantities β_* and a_0 by $-\beta^*$ and $-a_0$, respectively. □

Now, in view of the interrelation between $\Lambda(\alpha)$ and $\mathcal{L}(a)$, $a = (\alpha + 1)/\alpha$, stated in Corollary 22.1 and the analytic prolongability with respect to λ, the semigroup character of $\{\mathcal{L}(a)^{\lambda}\}_{\lambda}$ implies that of $\{\Lambda(\alpha)^{\lambda}\}_{\lambda}$; namely,

$$\Lambda(\alpha)^{\lambda} = \Lambda(\alpha)^{\mu}\, \Lambda(\alpha)^{\lambda-\mu} \quad \text{for } 0 < \mu < \lambda.$$

Hence, by repeated application of Theorem 20.2, we see that

$\Lambda(\alpha)^{\lambda} f(z) > \beta_{*}$ implies

$$\text{Re } \Lambda(\alpha)^{\mu} f(z) > \beta_{*} + (a_0 - \beta_{*}) \Psi(\lambda - \mu, \alpha),$$

$$\begin{aligned}
&\text{Re } f(z) \\
&> \beta_{*} + (a_0 - \beta_{*}) \Psi(\lambda - \mu, \alpha) \\
&\quad + (a_0 - (\beta_{*} + (a_0 - \beta_{*}) \Psi(\lambda - \mu, \alpha)) \Psi(\mu, \alpha) \\
&= \beta_{*} + (a_0 - \beta_{*})(\Psi(\lambda - \mu, \alpha) + (1 - \Psi(\lambda - \mu, \alpha)) \Psi(\mu, \alpha).
\end{aligned}$$

By comparing the lower bound

$$B_{*}^{(1)} = \beta_{*} + (a_0 - \beta_{*}) \Psi(\lambda, \alpha)$$

for Re $f(z)$ given in Theorem 22.2 with the bound $B_{*}^{(2)}$ obtained in the last expression, their difference becomes

$$\begin{aligned}
&B_{*}^{(1)} - B_{*}^{(2)} \\
&= (a_0 - \beta_{*})((1 - \Psi(\mu, \alpha))(1 - \Psi(\lambda - \mu, \alpha) - (1 - \Psi(\lambda, \alpha)).
\end{aligned}$$

It may be directly shown that the logarithm of the quantity

$$1 - \Psi(\lambda, \alpha) = \frac{1}{\alpha^{\lambda} \Gamma(\lambda)} \int_I t^{1/\alpha - 1} \left(\log \frac{1}{t}\right)^{\lambda - 1} \frac{2t}{1 + t} \, dt$$

possesses the subadditivity with respect to λ, i.e.,

$$1 - \Psi(\lambda, \alpha) < (1 - \Psi(\mu, \alpha))(1 - \Psi(\lambda - \mu, \alpha)), \quad 0 < \mu < \lambda,$$

whence follows $B_{*}^{(1)} > B_{*}^{(2)}$ verifying that the estimate based on Theorem 22.2 at one stretch is better than the estimate obtained by repeated steps. Concerning the upper bound given in Corollary 22.2, the circumstances are quite similar.

We mention some remarks to the bounds of distortion. The quantity $\Psi(\lambda, \alpha)$ is nothing but the quantity $\phi(\lambda, 1/\alpha)$ dealt with in § 18. It can be expressed in series form

$$\Psi(\lambda, \alpha) = 1 + \frac{1}{\alpha^{\lambda}} \sum_{\nu=2}^{\infty} \frac{(-1)^{\nu}}{(\nu + \alpha^{-1} - 1)^{\lambda}}$$

and if, in particular, $\alpha^{-1} = k$ is a positive integer, Ψ is alternatively expressible in the form

$$\Psi\left(\lambda, \frac{1}{k}\right)$$

$$= 1 + 2(-1)^{k-1} k^{\lambda}\left(\left(1 - 2^{1-\lambda}\right)\zeta(\lambda) + \sum_{\kappa=1}^{k} \frac{(-1)^{\kappa}}{\kappa^{\lambda}}\right),$$

ζ denoting Riemann zeta function for which the value of $(1 - 2^{1-\lambda})\zeta(\lambda)$ at $\lambda = 1$ is to be understood its limit value log 2.

It has been also noted that $\Psi(\lambda, \alpha)$ possesses monotonicity with respect to each argument. In fact, for any fixed $\alpha > 0$ or fixed $\lambda > 0$, $\Psi(\lambda, \alpha)$ increases strictly from 0 to 1 when λ or α increases from 0 to ∞, respectively.

Now, we deal with the sequence $\{\Psi(\lambda, 1/k)\}_{k=1}^{\infty}$ for any fixed λ. In view of the above-stated expression for ψ, it follows readily that it satisfies the recurrence relation

$$\psi\left(\lambda, \frac{1}{k+1}\right) = \left(1 + \frac{1}{k+1}\right)^{\lambda} - 1 - \left(1 + \frac{1}{k}\right)^{\lambda} \psi\left(\lambda, \frac{1}{k}\right).$$

For $\lambda = 1$, we have

$$\psi(1, 1) = 2 \log 2 - 1,$$

while for an even integer $\lambda = 2m$ $(m = 1, 2, \ldots)$, we have

$$\psi(2m, 1) = 2\left(1 - 2^{1-2m}\right)\zeta(2m) - 1$$

$$= \left(1 - 2^{1-2m}\right)(2\pi)^{2m} \frac{(-1)^{m-1} B_{2m}}{(2m)!} - 1,$$

where B's denote the Bernoullian numbers defined by the generating function

$$\frac{t}{\exp t - 1} = \sum_{n=1}^{\infty} \frac{B_n}{n!} t^n.$$

In particular, for lower values of m, the numerical values of B's are given by

$$B_2 = \frac{1}{6}, \quad B_4 = -\frac{1}{30}, \quad B_6 = \frac{1}{42}, \quad B_8 = -\frac{1}{10},$$

$$B_{10} = \frac{5}{66}, \quad B_{12} = -\frac{691}{2730}, \quad B_{14} = \frac{7}{6}, \quad \ldots .$$

Hence, in view of the above-mentioned recurrence relation, the numerical values of $\Psi(\lambda, 1/k)$ with $\lambda = 1$ or $2m$ ($m = 1, 2, \ldots$) and $k = 1, 2, \ldots$ will be successively determined. For instance,

$$\Psi(1, 1) = 2\log 2 - 1 = 0.3862\ldots,$$

$$\Psi\left(1, \frac{1}{2}\right) = 3 - 4\log 2 = 0.2274\ldots,$$

$$\Psi\left(1, \frac{1}{3}\right) = 6\log 2 - 4 = 0.1588\ldots;$$

$$\Psi(2, 1) = \frac{\pi^2}{6} - 1 = 0.6449\ldots,$$

$$\Psi\left(2, \frac{1}{2}\right) = 7 - \frac{2\pi^2}{3} = 0.4202\ldots,$$

$$\Psi\left(2, \frac{1}{3}\right) = \frac{3\pi^2}{2} - \frac{29}{2} = 0.3044\ldots;$$

$$\Psi(4, 1) = \frac{7\pi^4}{360} - 1 = 0.8940\ldots,$$

$$\Psi\left(4, \frac{1}{2}\right) = 31 - \frac{14\pi^4}{45} = 0.6949\ldots,$$

$$\Psi\left(4, \frac{1}{3}\right) = \frac{567\pi^4}{360} - \frac{1223}{8} = 0.5443\ldots.$$

§ 23. Application of subordination

Now, we intend to generalize the distortion Theorem 21. 2. The background of the proof is based on a simple idea of subordination; to the present section, cf. Komatu [22].

Let D be a simply-connected domain laid on the w-plane which is th e image of E by a mapping $w = W(z)$ satisfying $W(0) = f(0)$. The Green function of the domain D with the pole at $W(0)$ is given by $\log |W^{-1}(w)|^{-1}$ and hence its level curve C_t of height $\log t^{-1}$ is expressed by $|W^{-1}(w)| = t$. The preimage of C_t is the concentric circumference $\{|z| = t\}$. The image of $\{|z| < t\}$ is the interior of C_t which will be denoted by D_t.

If $\Lambda(\alpha)^{\lambda} f(z) \subset D$, then it is subordinate to $W(z)$ and hence $\Lambda(\alpha)^{\lambda} f(zt) \subset D_t$ for any $z \in E$ and $t \in I$. Consequently, for any $z \in E$, the quantity

$$f(z) = \Lambda(\alpha)^{-\lambda} \Lambda(\alpha)^{\lambda} f(z)$$

$$= \frac{1}{\alpha^{\lambda}\Gamma(\lambda)} \int_I t^{1/\alpha-1}\left(\log \frac{1}{t}\right)^{\lambda-1} \Lambda(\alpha)^{\lambda} f(zt)\, dt$$

may be regarded as a weighted mean of $\Lambda(\alpha)^{\lambda} f(zt) \in D_t$ along I with a positive weight.

Accordingly, we can mention a theorem generalizing Theorem 22.2 together with its Corollary 22.2 as in the following manner.

THEOREM 23.1. *Let* $f \in \tilde{\mathcal{F}}$. *Let* D *be a simply - connected domain and* D *be the interior of the level curve of height* $\log t^{-1}$ *of its Green function with pole at* $f(0)$. *Then*

$$\Lambda(\alpha)^{\lambda} f(E) \subset D, \quad \alpha > 0 \text{ and } \lambda > 0$$

implies

$$\Lambda(\alpha)^{\lambda} f(\{|z| < t\}) \subset D_t .$$

In particular, if D *is symmetric with respect to the horizontal line* $\{\operatorname{Im} w = \operatorname{Im} f(0)\}$, *then* $\operatorname{Re} \Lambda(\alpha)^{\lambda} f(E) \subset D$ *implies*

$$\Omega_* < \operatorname{Re} f(z) < \Omega^*$$

where Ω_* *and* Ω^* *are given by*

$$\left.\begin{matrix} \Omega_* \\ \\ \Omega^* \end{matrix}\right\} = \left\{\begin{matrix} \inf \\ \\ \sup \end{matrix}\right. \quad \frac{1}{\alpha^{\lambda}\Gamma(\lambda)} \int_I t^{1/\alpha - 1}\left(\log\frac{1}{t}\right)^{\lambda-1} \operatorname{Re} W(zt)\, dt ,$$

W *being a function mapping* E *onto* D *which satisfies* $W(0) = f(0)$; *here* inf *and* sup *are concerned in* z *lying on the circumference* $\{|z| = 1\}$.

Proof. The first part of the assertion follows readily from the subordination principle. For the second particular case, we have only to notice that the symmetry of D with respect to the line $\{\operatorname{Im} w = \operatorname{Im} f(0)\}$ implies that of D_t for any $t \in I$. □

As an example of Theorem 23.1, we mention here a result obtained by applying it to the case where D is a parallel strip. It relates to a result in Komatu [12].

COROLLARY 23.1. *If, in Theorem* 23.1, D *is a parallel strip* $\{\beta_* < \operatorname{Re} w < \beta^*\}$, *then the bounds of* $f(z)$ *are given by*

$$\left.\begin{matrix}\Omega_* \\ \Omega^*\end{matrix}\right\} = \frac{1}{\alpha^{\lambda}\Gamma(\lambda)} \int_I t^{1/\alpha-1}\left(\log\frac{1}{t}\right)^{\lambda-1} W(\mp t)\,dt,$$

W *denoting the function mapping* E *onto* D *with* $W(0) = a_0 + ib_0$:

$$W(z) = \frac{2(\beta^* - \beta_*)}{\pi} \arctan\frac{z+\tau}{1+\tau z} + \frac{\beta^* + \beta_*}{2} + ib_0$$

where

$$\tau = -\tan\left(\frac{\pi}{4}\,\frac{\beta^* + \beta_* - 2a_0}{\beta^* - \beta_*}\right).$$

The extremal function $\hat{f}$ *is of the form*

$$\hat{f}(z) = \Lambda(\alpha)^{\lambda} W(z) \quad \textit{with} \quad a_0 = \hat{a}_0,\ b_0 = \hat{b}_0 .$$

Its respective lower and upper bounds are attained at boundary points $z = -1$ *and* $z = +1$, *respectively, where* $\hat{f}$ *behaves holomorphically.*

Proof. Since the image strip $\{\beta_* < \operatorname{Re} w < \beta^*\}$ of E by the mapping $w = W(z)$ is convex, the image $W(\{|z| < t\})$ is so also for any $t \in I$. Consequently, in view of the symmetry with respect to the line $\{\operatorname{Im} w = b_0\}$, the minimum and maximum of $\operatorname{Re} W(z)$ as well as of $\operatorname{Re} A(\alpha)^{\lambda} W(z)$ on the circumference $\{|z| = t\}$ are attained at $z = -t$ and $z = +t$, respectively. Hence, in view of Theorem 23.1, the assertion follows. □

The corollary just stated may be regarded as an improvement of Theorem 22.2 together with its Corollary 22.2. As cared before, Corollary 23.1 reduces to Theorem 22.2 and Corollary 23.2 as the limit cases of $\beta^* \to +\infty$ and $\beta_* \to -\infty$, respectively.

It is shown that if we retain β_* and observe the limit case of $\beta^* \to +\infty$, the lower bound for $\operatorname{Re} f(z)$ given in Corollary 23.1 reduces to that given in Theorem 23.2. In fact, it is verified that as $\beta^* \to +\infty$, we obtain

$$\tau = -1 + \frac{\pi(a_0 - \beta_*)}{\beta^*} + o\left(\frac{1}{\beta^*}\right),$$

$$\arctan \frac{\tau - t}{1 - \tau t} = -\frac{\pi}{4} + \frac{\pi(a_0 - \beta_*)}{2\beta^*} \frac{1 - t}{1 + t} + o\left(\frac{1}{\beta^*}\right),$$

every o-notation being uniform, and hence

$$\Omega_* = \frac{2(\beta^* - \beta_*)}{\pi} \frac{1}{\alpha^\lambda \Gamma(\lambda)} \int_I t^{1/\alpha - 1}\left(\log \frac{1}{t}\right)^{\lambda - 1} \arctan \frac{\tau - t}{1 - \tau t}\, dt + \frac{\beta^* + \beta_*}{2}$$

$$= \beta_* + (a_0 - \beta_*) \Psi(\lambda, \alpha) + o(1).$$

Similarly, as $\beta_* \to -\infty$, we obtain

$$\Omega^* = \beta^* - (\beta^* - a_0) \Psi(\lambda, \alpha) + o(1).$$

Chapter 6. Distortions on Miscellaneous Functionals

§ 24. Oscillation

Among miscellaneous functionals, we shall deal in the present section with the distortions on oscillation of the real part of analytic functions holomorphic in the unit disk E along a circumference about the origin; cf. Komatu [13].

For an analytic function g holomorphic in E, the oscillation of $\operatorname{Re} g(\zeta)$ on a concentric circumference $\{|\zeta| = r\}$ is denoted by

$$\operatorname*{osc}_{|\zeta|=r} \operatorname{Re} g(\zeta) := \max_{|\zeta|=r} \operatorname{Re} g(\zeta) - \min_{|\zeta|=r} \operatorname{Re} g(\zeta).$$

By the maximum-minimum principle, we may replace the affixed set $\{|\zeta| = r\}$ by $\{|\zeta| \leq r\}$. Moreover, we may also write $\{|\zeta| < r\}$ instead of $\{|\zeta| \leqq r\}$, provided we replace max and min by sup and inf, respectively. In particular, we have

$$\operatorname*{osc}_{|\zeta|\leqq r} \operatorname{Re} g(\zeta) = \sup_{|\zeta|<r} \operatorname{Re} g(\zeta) - \inf_{|\zeta|<r} \operatorname{Re} g(\zeta).$$

This form applies even for $r = 1$.

We begin by referring to formulas of Neumann [1] which were restated by Koebe [1].

These will serve well as basic lemmas in the following lines.

LEMMA 24.1. *Let* g *be an analytic function holomorphic in* E *and normalized by* $g(0) = 0$ *at the origin. If* $\operatorname{Re} g \leqq \Delta/2$ *in* E, *then*

$$\operatorname*{osc}_{|\zeta|=r} \operatorname{Re} g(\zeta) \leqq \frac{4\Delta}{\pi} \arctan r$$

holds for $0 \leqq r < 1$. *The extremal function for* $r \in (0, 1)$ *is given only by* $g^{*}(\varepsilon\zeta)$ *with* $|\varepsilon| = 1$ *where*

$$g^{*}(\zeta) = \frac{2\Delta}{\pi} \arctan \zeta$$

is a function which maps E *onto the parallel strip* $\{|\operatorname{Re} \omega| < \Delta/2\}$.

Proof. The parallel strip $\{|\operatorname{Re} w| < \Delta/2\}$ is mapped by

$$w = \frac{4\Delta}{\pi} \arctan \zeta$$

onto $\{|\zeta| < 1\}$ in such a manner that $\zeta = 0$ corresponds to $W = 0$ and the concentric disk $\{|w| \leqq r\}$ corresponds to the parallel strip $\{|\operatorname{Re} w| \leqq r\Delta/2\}$. By means of the theorem Schwarz [1], the point $g(z)$ is contained in this strip, and hence

$$\operatorname*{osc}_{|\zeta|=r} \operatorname{Re} g(\zeta) \leqq \frac{4\Delta}{\pi} \arctan r .$$

That the extremal function is g^* as stated in the Lemma is evident. □

LEMMA 24.2. *Let g be an analytic function holomorphic in E and normalized by $g(0) = 0$. If it satisfies*

$$-a < \operatorname{Re} g(\zeta) < b \qquad (a, b > 0)$$

in E, then the distortion inequality

$$\frac{2(b+a)}{\pi} \arctan \frac{\tau - |\zeta|}{1 - \tau|\zeta|} + \frac{b-a}{2}$$

$$\leqq \operatorname{Re} g(\zeta) \leqq \frac{2(b+a)}{\pi} \arctan \frac{\tau + |\zeta|}{1 + \tau|\zeta|} + \frac{b-a}{2}$$

holds for $\zeta \in E$ where τ is defined by

$$\tau = -\tan\left(\frac{\pi}{4} \frac{b-a}{b+a}\right).$$

The extremal function for $\{0 < |\zeta| < 1\}$ is given only by $g^{**}(\varepsilon\zeta)$ with $|\varepsilon| = 1$ where g^{**} is a function mapping E onto the parallel strip $\{-a < \operatorname{Re} \omega < b\}$:

$$g^{**}(\zeta) = \frac{2(b+a)}{\pi} \arctan \frac{\tau + \zeta}{1 + \tau\zeta} + \frac{b-a}{2}.$$

Proof. The function

$$w = \frac{2(b+a)}{\pi} \arctan \frac{\tau + \zeta}{1 + \tau\zeta} + \frac{b-a}{2}$$

maps E onto the parallel strip $\{-a < \operatorname{Re} w < b\}$ such that the point $\tau = 0$ corresponds to the origin $w = 0$. By means of Schwarz theorem the disk $\{|\zeta| \leqq r\}$ is mapped onto the parallel strip mentioned in Lemma 24.1. The assertion with respect to the extremal function is readily verified. □

Let $\mathcal{F}$ be, as usual, the class of analytic function f holomorphic in E and normalized by $f(0) = f'(0) - 1 = 0$. And, let an operator $\mathcal{L}: f \mapsto \mathcal{L}f$ be defined on $\mathcal{F}$ by the integral representation

$$\mathcal{L}f(z) = \int_I \frac{f(zt)}{z}\, d\sigma(t),$$

where σ is a probability measure supported on the unit interval $I = [0, 1]$.

We now consider the oscillation of $\mathrm{Re}(f(z)/z)$ along $\{|z| = r\}$ which will be denoted by

$$\Delta(r) := \operatorname*{osc}_{|z|=r} \mathrm{Re}\,\frac{f(z)}{z}.$$

If we make use of the definition

$$\left.\begin{matrix} h(r) \\ \\ H(r) \end{matrix}\right\} = \begin{matrix} \min \\ \\ \max \\ {\scriptstyle |z|=r} \end{matrix}\ \mathrm{Re}\,\frac{f(z)}{z},$$

we may also write

$$\Delta(r) = H(r) - h(r).$$

For the sake of brevity, we shall use here also an abbreviated notation:

$$\hat{f}(z) = \mathcal{L} f(z) = \int_I \frac{f(zt)}{t} d\sigma(t)$$

and, similarly, we attach the symbol ^ to every quantity after transformation by the operator $\mathcal{L}$; for instance, $\hat{\Delta} = \mathcal{L}\Delta$, similarly as $\hat{H}$ or $\hat{h}$. In particular, we have

$$\hat{\Delta}(r) = \hat{H}(r) - \hat{h}(r).$$

The monotonicity of $\mathcal{L}$ for Δ, namely,

$$\hat{\Delta}(r) \leqq \Delta(r),$$

is readily seen. More precisely, we have the following theorem:

THEOREM 24.1. *We have*

$$\hat{\Delta}(r) \leqq \frac{4}{\pi} \Delta(r) \int_I \arctan t \, d\sigma(t).$$

In particular, the monotonicity $\hat{\Delta}(r) \leqq \Delta(r)$ *always holds. The equality sign in the estimation appears if and only if* σ *is the measure concentrated at* 1 *unless* $f(z) = z$.

Proof. For any fixed z with $|z| = r < 1$, the quantity $F(z\tau) = f(z\tau)/z\tau - 1$ can be regarded as a function of complex variable τ which is holomorphic on $\{|\tau| \leqq 1\}$. It satisfies $F(0) = f'(0) - 1 = 0$ and

$$\operatorname*{osc}_{|\tau|=1} \operatorname{Re} F(z\tau) = \operatorname*{osc}_{|\zeta|=r} \operatorname{Re} F(\zeta) = \Delta(r).$$

In view of Lemma 24.1 applied to $F(z\tau)$ as function of τ, we get for $t \in I$

$$\operatorname*{osc}_{|\tau|=t} \operatorname{Re} F(z\tau) \leqq \frac{4}{\pi} \Delta(r) \arctan t .$$

Consequently, the definition of $\mathcal{L}$ yields

$$\hat{\Delta}(r) = \operatorname*{osc}_{|z|=r} \operatorname{Re}\left(\frac{\hat{f}(z)}{z} - 1 \right)$$

$$= \operatorname*{osc}_{|z|=r} \int_I \operatorname{Re}\left(\frac{f(zt)}{zt} - 1\right) d\sigma(t)$$

$$\leqq \int_I \operatorname*{osc}_{|\tau|=t} \operatorname{Re} F(z\tau)\, d\sigma(t)$$

$$\leqq \frac{4\Delta(r)}{\pi} \int_I \arctan t \; d\sigma(t).$$

Since $\arctan t \leqq \pi/4$ in I, the latter part follows readily.

□

Now, by making use of Lemma 24.2, we can derive the monotonicity of $\mathcal{L}$ in more concrete form. For that purpose we prepare the following Theorem:

THEOREM 24.2. *For any* $f \in \mathcal{J}$ *and* $\delta > 0$

$$\frac{H_\lambda(r) - h_\lambda(r)}{2}\varphi_{\lambda,\delta}(r) + \frac{H_\lambda(r) + h_\lambda(r)}{2}$$

$$\leq h_{\lambda+\delta}(r) \leq H_{\lambda+\delta}(r)$$

$$\leq \frac{H_\lambda(r) - h_\lambda(r)}{2}\phi_{\lambda,\delta}(r) + \frac{H_\lambda(r) + h_\lambda(r)}{2}$$

where $\varphi_{\lambda,\delta}$ *and* $\phi_{\varphi,\delta}$ *are defined by*

$$\left.\begin{matrix}\varphi_{\lambda,\delta}(r)\\ \\ \phi_{\lambda,\delta}(r)\end{matrix}\right\} = \frac{4}{\pi}\int_I \arctan\frac{T_\lambda(r) \mp t}{1 \mp T_\lambda(r)t}$$

with

$$T_\lambda(r) = -\tan\left(\frac{\pi}{4}\,\frac{H_\lambda(r) + h_\lambda(r) - 2}{H_\lambda(r) - h_\lambda(r)}\right).$$

Proof. For any $f \in \mathcal{J}$ and any fixed z with $|z| = r < 1$, the quantity $F(zt) = f(zt)/zt - 1$ may be regarded as a function of a complex variable t holomorphic on $\{|t| \leq 1\}$. It satisfies $F(0) = f'(0) - 1 = 0$ and

$$-(1 - h(r)) \leq \operatorname{Re} F(zt) \leq H(r) - 1$$

for $\{|t| \leq 1\}$ and, in particular, for $\{t \in I\}$. Noting that

$$f_{\lambda+\delta}(z) = \mathcal{L}^\lambda f_\lambda(z)\int_I \frac{f_\lambda(zt)}{t}\,d\sigma_\delta(t),$$

in view of the Lemma 24.2 applied to $F(zt)$ as function of t, we get for $t \in I$

$$\frac{2(H(r)-h(r))}{\pi}\arctan\frac{T(r)-t}{1+T(r)t}+\frac{H(r)+h(r)-2}{2}$$

$$\leqq \operatorname{Re} F(zt)$$

$$\leqq \frac{2(H(r)-h(r))}{\pi}\arctan\frac{T(r)+t}{1+T(r)t}+\frac{H(r)+h(r)-2}{2}.$$

Applying this inequality for H_λ, h_λ instead of H, h and making attention to

$$\operatorname{Re}\frac{f_{\lambda+\delta}(z)}{z}-1=\operatorname{Re}\int_I\left(\frac{f_\lambda(zt)}{zt}-1\right)d\sigma_\delta(t)$$
$$=\int_I \operatorname{Re} F(zt)\,d\sigma_\delta(t),$$

we obtain the desired result. □

While the proof of Theorem 24.1 given above has been based directly on Lemma 24.1, the assertion can also be derived from Theorem 24.2. In fact, we have

$$\hat{\Delta}(r)$$

$$\leqq \frac{2\Delta(r)}{\pi}\int_I\left(\arctan\frac{T(r)+t}{1+T(r)t}-\arctan\frac{T(r)-t}{1-T(r)t}\right)d\sigma(t)$$

$$=\frac{2\Delta(r)}{\pi}\int_I \arctan\left(\frac{1-T(r)^2}{1+T(r)^2}\,\frac{2t}{1-t^2}\right)d\sigma(t).$$

In view of $h \leqq 1 \leqq H$ we have $-1 \leqq T \leqq 1$ and hence

$$\hat{\Delta}(r) \leqq \frac{2\Delta(r)}{\pi}\int_I \arctan\frac{2t}{1-t^2}\,d\sigma(t)$$

$$= \frac{4\Delta(r)}{\pi}\int_I \arctan t\,d\sigma(t).$$

We notice here that the estimation given in Theorem 24.2 remains valid even when h and H are replaced by any lower and upper bounds of $\mathrm{Re}(f(z)/z)$, respectively, instead of its minimum and maximum. Accordingly, if we would use h alone without referring to H, the estimation in Theorem 24.2 seems to reduce to Theorem 18.1. For that purpose, we have only to consider the limit as $H \to \infty$. Then, we get in turn

$$T = -\tan\left(\frac{\pi}{4}\,\frac{H+h-2}{H-h}\right) = -1 + \frac{\pi(1-h)}{H} + o\left(\frac{1}{H}\right),$$

$$\arctan\frac{T-t}{1-Tt} = -\frac{\pi}{4} + \frac{\pi(1-h)}{2H}\,\frac{1-t}{1+t} + o\left(\frac{1}{H}\right),$$

whence really follows

$$\frac{2(H-h)}{\pi}\int_I \arctan\frac{T-t}{1-Tt}\,d\sigma(t) + \frac{H+h}{2}$$

$$= h + (1-h)\int_I \frac{1-t}{1+t}\,d\sigma(t) + o(1).$$

Similarly, if we would retain H alone without referring to h, we obtain for the limit as $h \to -\infty$ the corresponding result on H.

Now, let the oscillation of $\mathrm{Re}((\mathcal{L}^{\lambda} f(z))/z)$ for $f \in \mathcal{J}$ on $\{|z| = r\}$ be denoted by

$$\Delta_\lambda(r) = \operatorname*{osc}_{|z|=r} \operatorname{Re} \frac{\mathcal{L}^\lambda f(z)}{z} .$$

In the following lines, we restrict ourselves to the particular case where the probability measure is generated by $\sigma(t) = t$. Then, as a Corollary of Theorem 24.2, we have the following result.

COROLLARY 24.1. *For any* $f \in \mathcal{J}$ *and* $\delta > 0$ *the inequality*

$$\Delta_{\lambda+\delta}(r) \leqq Q(\delta)\Delta_\lambda(r)$$

holds where the factor Q *is given by*

$$Q(\delta) = \frac{4}{\pi}\frac{1}{\Gamma(\delta)}\int_I \arctan t \left(\log\frac{1}{t}\right)^{\delta-1} dt$$

which satisfies $0 < Q(\delta) < 1$.

Proof. We have only to notice the additivity of $\{\mathcal{L}^\lambda\}$:

$$\mathcal{L}^{\lambda+\delta} f(z) = \mathcal{L}^\delta \mathcal{L}^\lambda f(z) = \frac{1}{\Gamma(\delta)}\int_I \frac{\mathcal{L}^\lambda f(zt)}{t}\left(\log\frac{1}{t}\right)^{\delta-1} dt.$$

By applying Theorem 24.1 to $\mathcal{L}^\lambda f$ instead of f and specializing correspondingly the measure under consideration, we readily obtain the desired estimation. Since $\arctan t$ increases strictly in I, the inequality $0 < Q(\delta) < 1$ follows. □

By the way, we notice here that the value $Q(0)$ is equal to 1. In fact, by means of integration by parts, we get for $\delta > 0$

$$\mathcal{Q}(\delta) = \frac{4}{\pi} \frac{1}{\Gamma(\delta+1)} \int_I \frac{d}{dt}(t \arctan t) \left(\log \frac{1}{t}\right)^{\delta} dt,$$

whence readily follows

$$\mathcal{Q}(0) \equiv \lim_{\delta \to +0} \mathcal{Q}(\delta) = \frac{4}{\pi} \int_I \frac{d}{dt}(t \arctan t)\, dt = 1.$$

We now derive a differential inequality for Δ_λ concerning the right-hand derivative with respect to λ.

THEOREM 24.3. *For any $f \in \mathcal{F}$, the oscillation Δ_λ satisfies*

$$\frac{\partial}{\partial \lambda} \Delta_\lambda(r) \leqq \mathcal{Q}'(0) \Delta_\lambda(r),$$

with a numerical factor given by

$$\mathcal{Q}'(0) = \frac{4}{\pi} \int_I \log \log \frac{1}{t} \frac{d}{dt}(t \arctan t)\, dt + C,$$

C being the Euler constant.

Proof. In view of $\mathcal{Q}(0) = 1$ as noticed above, the inequality stated in Corollary 24.1 leads to

$$\frac{\delta_{\lambda+\delta}(r) - \delta_\lambda(r)}{\delta} \leqq \frac{\mathcal{Q}(\delta) - \mathcal{Q}(0)}{\delta} \Delta_\lambda(r).$$

whence follows the desired differential inequality after $\delta \to +0$. Now, by differentiating the expression for $\mathcal{Q}(\delta)$ obtained after integration by parts, we have

$$\mathcal{Q}'(\delta) = \frac{4}{\pi}\int_I \left(\frac{1}{\Gamma(\delta+1)}\log\log\frac{1}{t} - \frac{\Gamma'(\delta+1)}{\Gamma(\delta+1)^2}\right)\left(\log\frac{1}{t}\right)^{\delta}\frac{d}{dt}(t\arctan t)\,dt\,.$$

Thus we get

$$\mathcal{Q}'(0) = \frac{4}{\pi}\int_I \left(\log\log\frac{1}{t} - \Gamma'(1)\right)\frac{d}{dt}(t\arctan t)\,dt\,,$$

whence follows the desired expression for $\mathcal{Q}'(0)$ by remembering $\Gamma'(1) = -C$. □

From the differential inequality stated in Theorem 24.3 we can derive an ordinary finite inequality. In fact, by dividing its both members by $\Delta_\lambda(r)$ followed by integrating with respect to λ, we obtain

$$\log\frac{\Delta_\delta(r)}{\Delta(r)} \leq \mathcal{Q}'(0)\delta, \quad \text{i. e.,} \quad \Delta_\delta(r) \leq \Delta(r)\exp(\mathcal{Q}'(0)\delta).$$

Certain evaluation concerning $\mathcal{Q}$ will be dealt with more closely. By expanding $(d/dt)(t\arctan t)$ in power series and integrating termwise the expression for $\mathcal{Q}$ given above, we get

$$\mathcal{Q}(\delta) = \frac{4}{\pi}\sum_{j=1}^{\infty}(-1)^{j-1}\frac{2j}{2j-1}\frac{1}{\Gamma(\delta+1)}\int_I t^{2j-1}\left(\log\frac{1}{t}\right)^{\delta}dt.$$

Since the integral contained in the summand is equal to $\Gamma(\delta + 1)/(2j)^{\delta+1}$, we obtain the series form for $\mathcal{Q}$:

$$Q(\delta) = \frac{4}{\pi} \sum_{j=1}^{\infty} (-1)^{j-1} \frac{1}{2j-1} \frac{1}{(2j)^{\delta}}.$$

Hence we get by termwise differentiation

$$Q'(\delta) = -\frac{4}{\pi} \sum_{j=1}^{\infty} (-1)^{j-1} \frac{1}{2j-1} \frac{\log 2j}{(2j)^{\delta}}$$

and, in particular,

$$Q'(0) = -\frac{4}{\pi} \sum_{j=1}^{\infty} \frac{\log 2j}{2j-1}.$$

By the way, we see that the last relation can also be derived directly from that given in Theorem 24.3. In fact, we get

$$Q'(0) = \frac{4}{\pi} \sum_{j=1}^{\infty} (-1)^{j-1} \frac{2j}{2j-1} \int_{I} t^{2j-1} \log\log \frac{1}{t}\, dt + C$$

$$= \frac{4}{\pi} \sum_{j=1}^{\infty} (-1)^{j-1} \frac{2j}{2j-1} \left(- \frac{\log 2j + C}{2j} \right) + C$$

$$= -\frac{4}{\pi} \sum_{j=1}^{\infty} (-1)^{j-1} \frac{\log 2j}{2j-1}.$$

It seems from this calculation thet the Euler constant has intermixed in the above expression rather apparently.

As seen above, $Q'(\delta)$ with $\delta \geqq 0$ is expressible in terms of alternating series with terms descending in absolute value and tending to zero as $j \to \infty$. Hence, $Q'(\delta)$ remains negative so that $Q(\delta)$ decreases strictly with respect to δ.

We notice that for $n = 0, 1, 2, \ldots$ we have

$$Q(n) - Q(n+1) = \frac{4}{\pi} \sum_{j=1}^{\infty} \frac{(-1)^{j-1}}{(2j)^{n+1}}.$$

In view of $Q(0)$ we get, in particular,

$$Q(1) = 1 - \frac{2\log 2}{\pi}, \quad Q(2) = 1 - \frac{2\log 2}{\pi} - \frac{\pi}{12}.$$

For odd integer $n = 2m - 1$ $(m = 1, 2, \ldots)$ each difference is expressible in terms of Bernoullian number in a compact form

$$Q(2m-1) - Q(2m) = \frac{(-1)^{m-1}}{2^{2m-1}} \frac{2^{2m-1} - 1}{(2m)!} B_{2m} \pi^{2m-1},$$

but for even integer n there is no such analogue.

Finally, we observe the numerical value of $Q'(0)$. The alternating series

$$s = \sum_{j=0}^{\infty} (-1)^j v_j, \quad v_j = \frac{\log(j+2)}{2j+3}$$

contained in the expression

$$Q'(0) = -\log 2 + \frac{4}{\pi} \sum_{j=2}^{\infty} (-1)^j \frac{\log j}{2j-1} = -\log 2 + \frac{4}{\pi} s$$

converges surely but very slowly, so that it is unsuitable by itself, especially in practice. In order to accelerate the convergence velocity, we may apply a transformation. For that purpose, let the differences of the first and second orders of the sequence $\{v_j\}$ be denoteed by $\{D v_j\}$ and $\{D^2 v_j\}$:

$$D v_j = v_j - v_{j+1}, \quad D^2 v_j = D v_j - D v_{j+1} = v_j - 2 v_{j+1} + v_{j+2}.$$

Then the series is transformed into

$$s = \frac{1}{2} v_0 + \frac{1}{2} \sum_{j=0}^{\infty} (-1)^j D v_j$$

$$= \frac{1}{2} v_0 + \frac{1}{4} D v_0 + \frac{1}{4} \sum_{j=0}^{\infty} (-1)^j D^2 v_j.$$

If we put

$$v(x) = \frac{\log(x + 2)}{2x + 3}, \quad v(j) = v_j,$$

then it is shown b y means of slementary calculus that $v''(x) > 0$ for $x \geqq 2$ and $v'''(x) < 0$ for $x \geqq 4$. Hence we have

$$D^2 v_j = \int_I dx \int_I v''(j + x + y)\, dy > 0 \quad \text{for } j \geqq 2,$$

$$D^2 v_j - D^2 v_{j+1} \quad (= D^3 v_j)$$

$$= - \int_I dx \int_I dy \int_I v'''(j + x + y + z)\, dz > 0 \quad \text{for } j \geqq 4.$$

It is ensured directly, for instance numerically, that $D^2 v_2 >$ $D^2 v_3 > D^2 v_4$. Consequently, $\sum_{j=2}^{\infty} (-1)^j D^2 v_j$ is an alternating series with terms strictly decreasing in absolute value. Its convergence is considerably rapid compared with the original series $\sum (-1)^j v_j$. Thus we obtain, for instance, an approximation

$$Q'(0) = -\log 2 + \frac{4}{\pi}\left(\frac{1}{2} v_0 + \frac{1}{4} D v_0 + \frac{1}{4}(D^2 v_0 - D^2 v_1)\right) + R$$

$$= -\log 2 + \frac{4}{\pi}\left(\frac{23 \log 2}{42} - \frac{\log 3}{5} - \frac{\log 5}{36}\right) + R$$

with the remainder R satidfying

$$0 < R < \frac{1}{\pi} D^2 v_2 = \frac{1}{\pi}\left(\frac{\log 4}{7} - \frac{2 \log 5}{9} + \frac{\log 6}{11}\right).$$

Numerical computation shows

$$Q'(0) \doteqdot -0.5465, \qquad \frac{1}{\pi} D^2 v_2 \doteqdot 0.0112.$$

On the other hand, we obtained two estimates $Q(\delta)$ and $\exp(Q'(0)\delta)$ for Δ_δ / Δ. It seems previously plausible from the way of derivation that the first estimation is somewhat better than the second. Some examples show actually

$$Q(1) \doteqdot 0.5587, \qquad \exp Q'(0) \doteqdot 0.5790,$$

$$Q(2) \doteqdot 0.2969, \qquad \exp(2Q'(0)) \doteqdot 0.3352.$$

§ 25. Length and area

Let $P(\alpha)$ be the Carathéodory class of order $\alpha < 1$ consisting of analytic functions p which are holomorphic, $p(0) = 1$ and $\operatorname{Re} p(z) > \alpha$. We denote by $\mathcal{F}(\alpha)$ the class of functions f which are of the form

$$f(z) = z\,p(z)$$

with $p \in P(\alpha)$.

Observing the length and the area related to the image mapped by $p(z)$, we shall deal with their distortions under the effect of the operator $\mathcal{L}$.

We begin with the following Lemmas originally obtained by Rogosinski [1]; cf. also Komatu [9], [11].

LEMMA 25.1. *Let* $f(z)/z \in \mathcal{F}(\alpha)$ *and* $L(r) = L(r;\, f)$ *denote the length of the image - curve of* $\{|z| = r < 1\}$ *by the mapping* $w = f(z)/z$. *Then, it satisfies*

$$L(r) \leqq (1 - \alpha)\,\frac{4\pi r}{1 - r^2}.$$

The equality sign appears for any fixed r *with* $0 < r < 1$ *if and only if* $f(z)/z$ *is a linear function mapping* E *onto* $\{\operatorname{Re} w > \alpha\}$, *i. e.*,

$$f(z) = z\,\frac{1 + (1 - 2\alpha)\varepsilon z}{1 - \varepsilon z} \quad \textit{with}\ |\varepsilon| = 1.$$

Proof. From $f \in \mathcal{F}(\alpha)$, we get

$$\frac{1}{1-\alpha}\left(\frac{f(z)}{z}-\alpha\right) \in P(0).$$

Hence, by means of Herglotz representation, we obtain

$$p(z) \equiv \frac{f(z)}{z} = (1-\alpha)\int_{-\pi}^{\pi} \frac{e^{i\varphi}+z}{e^{i\varphi}-z}\, d\tau(\varphi) + \alpha$$

$$= 2(1-\alpha)\int_{-\pi}^{\pi} \frac{e^{i\varphi}}{e^{i\varphi}-z}\, d\tau(\varphi) - (1-2\alpha)$$

where τ is a probability measure supported on $(-\pi, \pi]$. Direct calculation yields

$$L(r) = r\int_{-\pi}^{\pi} |p'(re^{i\theta})|\, d\theta$$

$$\leqq 2(1-\alpha)r\int_{-\pi}^{\pi} d\theta \int_{-\pi}^{\pi} \frac{1}{|e^{i\varphi}-re^{i\theta}|^2}\, d\tau(\varphi)$$

$$= 2\ (1-\alpha)r\int_{-\pi}^{\pi} d\tau(\varphi) \int_{-\pi}^{\pi} \frac{1}{|e^{i\varphi}-re^{i\theta}|^2}\, d\theta$$

$$= 2(1-\alpha)r\int_{-\pi}^{\pi} \frac{2\pi}{1-r^2}\, d\tau(\varphi)$$

$$= (1-\alpha)\frac{4\pi r}{1-r^2}.$$

Now, in order to attain the equality sign for an $r \in (0, 1)$,

It is necessary and sufficient that $e^{i\varphi}/(e^{i\varphi} - re^{i\theta})^2$ for every $\theta \in (-\pi, \pi)$ has the same argument as $\varphi \in (-\pi, \pi)$ coincident with $d\tau(\varphi) > 0$, and hence, the associated measure τ concentrates at a single point φ_0, say. Consequently, the extremal function is of the form

$$f(z) = z\left((1 - \alpha)\frac{1 + \varepsilon z}{1 - \varepsilon z}\right) = z\frac{1 + (1 - 2\alpha)\varepsilon z}{1 - \varepsilon z}$$

with $\varepsilon = e^{-i\varphi_0}$. □

Rogosinski [1] showed a Lemma for the class $P(\alpha)$ for which an alternative proof has been given in Komatu [1]; cf. also Komatu [11]:

LEMMA 25.2. *Let* $f \in \in \mathcal{J}(\alpha)$ *and* $A(r) = A(r; f)$ *denote the area of the image of the disk* $\{|z| < r < 1\}$ *by the mapping* $w = f(z)/z$ *where the area is to be calculated according to multiplicity. Then we have*

$$A(r) \leqq (1 - \alpha)^2 \frac{4\pi r^2}{(1 - r^2)^2}.$$

The bound is attained again by the same extremal function as given in Lemma 25.1.

Proof. In accordance with the notations used in the proof of the preceding Lemma 25.1, we obtain by directly calculating

$A(r)$

$$= \int_0^r d\rho \int_{-\pi}^{\pi} |p'(\rho e^{i\theta})|^2 d\theta$$

$$\leqq 4(1-\alpha)^2 \int_0^r \rho d\rho \int_{-\pi}^{\pi} d\theta \int_{-\pi}^{\pi} \frac{1}{|e^{i\varphi} - \rho e^{i\theta}|^4} d\tau(\varphi)$$

$$= 4(1-\alpha)^2 \int_{-\pi}^{\pi} d\tau(\varphi) \int_0^r \rho d\rho \int_{-\pi}^{\pi} \frac{1}{|e^{i\varphi} - \rho e^{i\theta}|^4} d\theta$$

$$= 4(1-\alpha)^2 \int_{-\pi}^{\pi} d\tau(\varphi) \int_0^r 2\pi \frac{\rho(1+\rho^2)}{(1-\rho^2)^3} d\rho$$

$$= 4(1-\alpha)^2 \int_{-\pi}^{\pi} \pi \frac{r^2}{(1-r^2)^2} d\tau(\varphi)$$

$$= (1-\alpha)^2 \frac{4\pi r^2}{(1-r^2)^2}.$$

The extremal function can be verified quite similarly as in Lemma 25.1. □

REMARK. While the above-mentioned proof of Lemma 24.2 has been based on the direct calculation, we could also give a proof by means of Lemma 25.1 together with the isoperimetric inequality, which is available also to non-univalent mapping; cf. Komatu [11]. In fact, we obtain simply

$$A(r) \leqq \frac{1}{4\pi} L(r)^2$$

$$\leqq \frac{1}{4\pi} \left((1-\alpha) \frac{4\pi r}{(1-r^2)^2} \right)^2$$

$$= (1-\alpha)^2 \frac{4\pi r^2}{(1-r^2)^2}.$$

The equality sign in the first inequality appears if and only if $p \in P$ is a linear function which maps the disk $\{|z| < r\}$ onto a disk and further E onto the half-plane $\{\mathrm{Re}\, w > \alpha\}$.

On the other hand, we here notice that Lemma 25.1 can be slightly generalized as in the following Lemma. For the sake of brevity, we suppose $\alpha = 0$; cf. Komatu [1].

LEMMA 25.3. *Let C be any measurable set contained in $(-\pi, \pi]$. Then, for any $p \in P(0)$, we have*

$$r\int_C |p'(re^{i\theta})|\, d\theta \leq \frac{8r}{1-r^2} \arctan\left(\frac{1+r}{1-r} \tan\frac{mC}{4}\right),$$

where mC denotes the measure of C. In case $mC > 0$ the equality sign holds for any fixed $r \in (0,1)$ if and only if C differs from an interval (u, v) by a set of measure zero and p is of the form $(1+\varepsilon z)/(1-\varepsilon z)$ with $\varepsilon = e^{-i(u+v)/2}$ or $e^{-i\varphi_0}$ for $v < u + 2\pi$ or $v = u + 2\pi$, respectively, φ_0 being an arbitrary real number.

Proof. We obtain quite similarly as in the proof of Lemma 24.1 the estimation

$$\int_C |p'(re^{i\theta})|\, d\theta \leq 2\int_C \frac{d\theta}{|1-re^{i\theta}|^2}.$$

Now, the integrand of the last integral, namely, the quantity $1/(1 - 2r\cos\theta + r^2)$ is an even function of θ which strictly decreases as θ increases from 0 to π. Hence, we get

$$\int_C \frac{d\theta}{|1 - re^{i\theta}|^2} \leqq 2\int_0^{mC/2} \frac{d\theta}{|1 - re^{i\theta}|^2}$$

$$= \frac{4}{1 - r^2} \arctan\left(\frac{1 + r}{1 - r} \tan \frac{mC}{4} \right).$$

The assertion about the extremal function is also verified similarly as in the Lemma 25.1. □

Corresponding to Lemma 25.3, we can derive as a consequence of itself an analogous estimation on the areal distortion:

LEMMA 25.4. *Let C be any measurable set contained in $(-\pi, \pi]$. Then, for any $p \in P(0)$, we have*

$$\int_{r_0}^{r} r\,dr \int_C |p'(re^{i\theta})|^2 d\theta$$

$$\leqq \Big[\frac{8r^2}{(1 - r^2)^2} \arctan\left(\gamma \frac{1 + r}{1 - r}\right) + \frac{r}{1 - r^2}\left(\gamma(1 - r) + \frac{1 + r}{\gamma}\right)$$

$$- \frac{(1 + \gamma^2)^2}{2\gamma^2} \arctan \frac{2\gamma r}{\gamma^2 + 1 + (\gamma^2 - 1)r} \Big]_{r_0}^{r}$$

$$(0 \leqq r_0 < r < 1)$$

where we put $\gamma = \tan(mC/4)$. *In the case of* $mC > 0$, *the extremal functions are characterized by the same condition as mentioned in the Lemma* 25.3.

Proof. Quite similarly as in the proof of Lemma 24.3, we get

$$\int_C |p'(re^{i\theta})\, d\theta| \leqq 4 \int_{r_0}^{r} \frac{d\theta}{|1 - re^{i\theta}|^4}$$

$$= \frac{4}{1 - r^2} \frac{d}{dr} \left(\frac{2r}{1 - r^2} \arctan\left(\gamma \frac{1 + r}{1 - r}\right)\right).$$

Integration with respect to r after mutiplied by r leads us readily to

$$\int_{r_0}^{r} r\, dr \int_C |p'(re^{i\theta})|^2\, d\theta$$

$$\leqq 4 \int_{r_0}^{r} \frac{r}{1 - r^2} \frac{d}{dr} \left(\frac{2r}{1 - r^2} \arctan\left(\gamma \frac{1 + r}{1 - r}\right)\right) dr$$

and actual evaluation of the last integral gives the desired result. The assertion on extremal functions is also evident. □

It is noted that the estimate in Lemma 25.4 just shown, expresses, of course, the area of the image-domain of the curvilinear quadrilateral $\{|\arg z| < mC/2,\ r_0 < |z| < r\}$ by the mapping $w = (1 + z)/(1 - z)$.

Here, it is noted, by the way, that the integral of the form

$$S_\lambda(r, \sigma) = \int_0^\sigma \frac{d\theta}{|1 - re^{i\theta}|^\lambda} = \int_0^\sigma \frac{d\theta}{(1 - 2r\cos\theta + r^2)^{\lambda/2}}$$

with any constant $\sigma \geqq 0$ can be evaluated in terms of elementary functions provided λ is an even integer. Here, we are interested in the case $\lambda > 0$. Actual calculation will show that there exists, in general, a recurrence formula

$$S_{\lambda+2}(r, \sigma) = \frac{2}{\lambda r^{\lambda/2-1}(1 - r)^2} \frac{d}{dr}(r^{\lambda/2} S_\lambda(r, \sigma)).$$

We may suppose $0 \leqq \sigma \leqq \pi$. Since we have, in particular,

$$S_2(r, \sigma) = \int_0^\sigma \frac{d\theta}{|1 - re^{i\theta})|^2} = \frac{2}{1 - r^2} \arctan\left(\frac{1 + r}{1 - r} \tan\frac{\sigma}{2}\right),$$

the quantity $S_\lambda(r, \sigma)$ with an even positive integer λ as suffix can be obtained from $S_2(r, \sigma)$ by means of repeated differentiation combined with elementary operations. We observe, for instance, the particular case $\sigma = \pi$. Then, we have $S_2(r, \pi) = \pi/(1 - r^2)$, and it is evident that $S_\lambda(r, \pi)$ with such a λ is a rational function of r^2. Moreover, it will be verified, for instance, by induction with respect to $\lambda/2$ that $S_\lambda(r, \pi)$ is expressed by the formula

$$S_\lambda(r, \pi) = \frac{\pi}{(1 - r^2)^{\lambda-1}} \sum_{j=0}^{\lambda/2-1} \binom{\lambda/2 - 1}{j} r^{2j}$$

provided λ is an even integer.

On the other hand, for such a value of λ, the quantity defined by

$$T_\lambda(r, \sigma) = \int_0^\sigma \left| \frac{1 + re^{i\theta}}{1 - re^{i\theta}} \right|^\lambda d\theta$$

$$= \int_0^\sigma \left(\frac{2(1 + r^2)}{1 - 2r\cos\theta + r^2} - 1 \right)^{\lambda/2} d\theta$$

is connected with $S_\lambda(r, \sigma)$ by the relation

$$T_\lambda(r, \sigma) = \sum_{\kappa=1}^{\lambda/2} (-1)^{\lambda/2-\kappa} \binom{\lambda/2}{\kappa} 2^\kappa (1 + r^2)^\kappa S_{2\kappa}(r, \sigma) - (-1)^{\lambda/2}.$$

In particular, the quantity $T_\lambda(r, \pi)$ with an even positive integer λ as suffix is a rational function of r^2 which can be explicitly written down.

Though the generalizations of Lemmata 25.1 and 25.2 established just now in the Lemmata 25.3 and 25.4 will be possible to apply to the subsequent theorems, we shall derive in the following lines only the distortion theorems about the length $L(r)$ and the area $A(r)$ for $f \in \mathcal{J}(\alpha)$ under the act of $\mathcal{L}$:

THEOREM 25.1. *Let $L(r) = L(r; f)$ denote the length of the image - curve of $\{|z| = r < 1\}$ by the mapping by*

$$w = \frac{f(z)}{z}$$

with $f \in \mathcal{F}$. *Then, after transforming by*

$$f \mapsto f_\lambda = \mathcal{L}^\lambda f: \quad f_\lambda(z) = \int_I \frac{f(zt)}{t}\, d\sigma_\lambda(t),$$

we have for $\lambda > 0$

$$L_\lambda(r) \leq \int_I L(rt)\, d\sigma_\lambda(t).$$

Proof. Let $p(z) = f(z)/z$ and $p_\lambda(z) = f_\lambda(z)/z$. Then, we get

$$p_\lambda(z) = \frac{f_\lambda(z)}{z} = \int_I \frac{f(zt)}{zt}\, d\sigma_\lambda(t) = \int_I p(zt)\, d\sigma_\lambda(t)$$

and consequently

$$\begin{aligned} L_\lambda(r) &= r\int_{-\pi}^{\pi} |p_\lambda'(re^{i\theta})|\, d\theta \\ &= r\int_{-\pi}^{\pi} d\theta \left| \int_I t\, p'(rte^{i\theta})\, d\sigma_\lambda(t) \right| \\ &\leq r\int_{-\pi}^{\pi} d\theta \int_I t\, |p'(rte^{i\theta})|\, d\sigma_\lambda(t) \\ &= \int_I d\sigma_\lambda(t)\; rt\int_{-\pi}^{\pi} |p'(rte^{i\theta})|\, d\theta \\ &= \int_I L(rt)\, d\sigma_\lambda(t). \end{aligned}$$

□

COROLLARY 25.1. *We have for* $f \in \mathcal{F}(\alpha)$

$$L_\lambda(r) \leqq (1-\alpha)4\pi r \int_I \frac{t}{1-r^2t^2}\, d\sigma_\lambda(t).$$

Proof. We only have to substitute the estimation into that for $L_\lambda(r)$ after applying to $L(rt)$. □

Corresponding distortions concerning the area-functional are similarly obtained:

THEOREM 25.2. *Under the similar notations as in Theorem* 25.1, *we have*

$$A_\lambda(r) \leqq \int_I A(rt)\, d\sigma_\lambda(t).$$

Proof. Direct calculation as in Theorem 25.2 shows

$$A_\lambda(r) = \int_0^r \rho\, d\rho \int_{-\pi}^{\pi} |p_\lambda'(\rho e^{i\theta})|^2 d\theta$$

$$= \int_0^r \rho\, d\rho \int_{-\pi}^{\pi} d\theta \left| \int_I tp'(\rho te^{i\theta})\, d\sigma_\lambda(t) \right|^2$$

and by virtue of Schwarz inequality

$$A_\lambda(r)$$

$$\leqq \int_0^r \rho\, d\rho \int_{-\pi}^{\pi} d\theta \int_I d\sigma_\lambda(t) \int_I t^2 |p'(\rho te^{i\theta})|^2 d\sigma_\lambda(t)$$

$$= \int_I d\sigma_\lambda(t) \int_0^r \rho t^2 d\rho \int_{-\pi}^{\pi} |p'(\rho te^{i\theta})|^2 d\theta$$

$$= \int_I d\sigma_\lambda(t) \int_0^{rt} \zeta \, d\zeta \int_{-\pi}^{\pi} |p'(\zeta e^{i\theta})|^2 d\theta$$

$$= \int_I A(rt)\, d\sigma_\lambda(t). \qquad \square$$

Similarly as Corollary 25.1, we obtain the following Corollary:

COROLLARY 25.2. *We have for* $f \in \mathcal{F}(\alpha)$

$$A_\lambda(r) \leqq (1-\alpha)^2 \, 4\pi r^2 \int_I \frac{t^2}{(1-r^2t^2)^2} d\sigma_\lambda(t).$$

Proof. Quite similarly to that of Corollary 25.1. □

If we apply the isoperimetric inequality directly to the pair $A_\lambda(r)$ and $L_\lambda(r)$, we shall be able to derive a more precise estimation from Corollary 25.1. In fact, we have the following result:

COROLLARY 25.3. *We have for* $f \in \mathcal{F}(\alpha)$

$$A_\lambda(r) \leqq (1-\alpha)^2 \, 4\pi r^2 \left(\int_I \frac{t}{1-r^2t^2} d\sigma_\lambda(t) \right)^2$$

By the way, we remark that there is always an inequality of inverse nature. In fact, in virtue of Schwarz inequality we can derive the following inequality:

$$\frac{1}{2\pi}\int_0^r \frac{L_\lambda(\rho)^2}{\rho}\, d\rho$$

$$= \frac{1}{2\pi}\int_0^r \rho \left(\int_{-\pi}^{\pi} |p_\lambda' \rho e^{i\theta}| \, d\theta\right)^2 d\rho$$

$$\leq \int_0^r \rho \, d\rho \int_{-\pi}^{\pi} |p_\lambda'(\rho e^{i\theta})|^2 d\theta$$

$$= A_\lambda(r).$$

§ 26. Evaluation of bounds

In the present section, we evaluate the bounds obtained in the preceding section for several subclasses; cf. Komatu [9].

First, we cosider the subclass $\mathscr{H} = P(1/2)$ of $\mathscr{F}^+ = P(0)$, which is characterized by

$$\operatorname{Re} \frac{f(z)}{z} > \frac{1}{2}.$$

This condition is equivalent to $2f(z) - z \in \mathscr{F}^+$. It is well-known that the familiar class $\mathscr{K}$ of normalized convex mappings is a proper subclass of $\mathscr{H}$; cf. Strohhäcker [1].

For this class, the length distortion is given by

$$L(r) \leq \frac{2\pi r}{1 - r^2}$$

with extremal function of the form

$$f(z) = \frac{z}{1 - \varepsilon z}.$$

This bound is equal to the half of that for $\mathcal{F}^+$.

We next consider the class $\mathcal{S}t$ of functions $f \in \mathcal{F}$ mapping E onto domains starlike with respect to the origin. As known, the condition for $f \in \mathcal{S}t$ is characterized by

$$z \frac{f'(z)}{f(z)} \in P(0);$$

cf. Komatu [9].

THEOREM 26.1. *Let $L(r)$ denote the length of the image-curve of $\{|z| = r < 1\}$ by the mapping*

$$w = z \frac{f'(z)}{f(z)} \in P(0).$$

Then, we have

$$L(r) \leqq \frac{8r}{(1 - r)^2(1 + r)} K\left(\frac{r^{1/2}}{1 + r}\right),$$

where K denotes the complete elliptic integral of the first kind with modulus $2r^{1/2}/(1 + r)$.

Proof. Herglotz representation applied to $zf'(z)/f(z) \in P(0)$ implies for $g(z) = f(z)/z$ the relation

$$g'(z) = \frac{zf'(z) - f(z)}{z^2} = \frac{f(z)}{z^2}\left(\frac{zf'(z)}{f(z)} - 1\right)$$

$$= \frac{f(z)}{z^2}\left(\int_{-\pi}^{\pi} \frac{e^{i\varphi}+z}{e^{i\varphi}-z}\, d\tau(\varphi) - 1\right)$$

$$= 2g(z)\int_{-\pi}^{\pi} \frac{1}{e^{i\varphi}-z}\, d\tau(\varphi)$$

with a probability measure τ supported by $(-\pi, \pi]$. Hence we get

$$L(r)$$

$$= r\int_{-\pi}^{\pi} |g'(re^{i\theta})|\, d\theta$$

$$= 2r\int_{-\pi}^{\pi} |g(re^{i\theta})|\, d\theta \left|\int_{-\pi}^{\pi} \frac{1}{e^{i\varphi}-re^{i\theta}}\, d\tau(\varphi)\right|$$

$$\leqq 2r\int_{-\pi}^{\pi} d\tau(\varphi)\int_{-\pi}^{\pi} \frac{|g(re^{i\theta})|}{|e^{i\varphi}-re^{i\theta}|}\, d\theta,$$

whence follows in view of the distortion theorem of Koebe [2],

$$L(r)$$

$$\leqq \frac{2r}{(1-r)^2}\int_{-\pi}^{\pi} \frac{d\theta}{|1-re^{i\theta}|}$$

$$= \frac{2r}{(1-r)^2}\, J(r), \quad \text{say.}$$

The explicit expression for $\mathcal{J}$ is obtained by means of direct elementary calculus. In fact,

$$\mathcal{J}(r) = 2\int_0^{\pi} \frac{d\theta}{(1 + r^2 - 2r\cos\theta)^{1/2}}$$

$$= 2\int_0^{\pi} \frac{d\theta}{((1 + r)^2 - 4r\cos^2(\theta/2))^{1/2}}$$

$$[\theta = \pi - 2\psi]$$

$$= 4\int_0^{\pi/2} \frac{d\psi}{((1 + r)^2 - 4r\sin^2\psi)^{1/2}}$$

$$= \frac{4}{1 + r}\int_0^{\pi/2} \frac{d\chi}{(1 - k^2\sin^2\chi)^{1/2}}$$

$$= \frac{4}{1 + r} K(k),$$

where $K(k)$ is the complete elliptic integral of the first kind with the modulus $k = 2r^{1/2}/(1 + r)$. □

Finally, we observe a particular case of $\mathcal{L}(a)$ generated by the measure $\sigma(t; a) = t^a$ with a parameter $a > 0$. In this case the measure σ_λ is given by

$$\sigma_\lambda(t; a) = \frac{a^\lambda}{\Gamma(\lambda)}\int_0^t \tau^{a-1}\left(\log\frac{1}{\tau}\right)^{\lambda-1} d\tau,$$

and hence the above-mentioned estimations are brought into more concrete forms.

THEOREM 26.2. *In the particular case* $\sigma(t\,;\,a) = t^{a}$ *with* $a > 0$, L_{λ} *for* $f \in \mathcal{F}(\alpha)$ *satisfies*

$$L_{\lambda}(r) \leqq (1-\alpha)\,\frac{4\pi a^{\lambda} r}{\Gamma(\lambda)} \int_I \frac{t^{a-1}}{1-r^2t^2}\left(\log\frac{1}{t}\right)^{\lambda-1} dt$$

$$=(1-\alpha)\,4\pi a^{\lambda} \sum_{\nu=1}^{\infty} \frac{r^{2\nu-1}}{(2\nu+a-2)^{\lambda}}\,.$$

Proof. In the estimation of Corollary 25.1, we have only to substitute the measure $\sigma_{\lambda}(t\,;\,a)$. The series form is implied in view of a known formula

$$\int_I t^{\kappa-1}\left(\log\frac{1}{t}\right)^{\lambda-1} = \frac{\Gamma(\lambda)}{\nu^{\lambda}}\,. \qquad \square$$

In particular, for $\lambda = 1$ we have

$$L_1(r) \leqq (1-\alpha)\,4\pi a\, r \int_I \frac{t^{a-1}}{1-r^2t^2}\, dt$$

$$= (1-\alpha)\,4\pi a \sum_{\nu=1}^{\infty} \frac{r^{2\nu-1}}{2\nu+a-2}\,.$$

While this bound is unbounded as $r \to 1-0$, $L_{\lambda}(r)$ remains bounded for every $\lambda > 1$. In fact, we have for $\lambda > 1$ always

$$L_\lambda(r) \leqq L_\lambda(1) = (1-\alpha)4\pi\left(\frac{a}{2}\right)^\lambda \zeta\left(\lambda;\frac{a-2}{2}\right) < \infty,$$

$\zeta(\lambda;\ c)$ denoting the generalized Riemann zeta function:

$$\zeta(\lambda;\ c) = \sum_{\nu=1}^{\infty} \frac{1}{(\lambda + c)^\lambda}.$$

Similar discussions can be made also made for the case of area distortion. In fact, we have the following theorem:

THEOREM 26.3. *In the case generated by the particular measure* $\sigma(t;\ a) = t^a$ *the estimation given in Cororally* 25.2 *becomes*

$$A_\lambda(r) \leqq (1-\alpha)^2 \frac{4\pi a^\lambda r^2}{\Gamma(\lambda)} \int_I \frac{t^a}{(1-r^2t^2)^2} \left(\log\frac{1}{t}\right)^{\lambda-1} dt$$

$$= (1-\alpha)^2\, 4\pi a^\lambda \sum_{\nu=1}^{\infty} \frac{\nu r^{2\nu}}{(2\nu + a - 1)^\lambda}.$$

Proof. Quite similarly to that of Theorem 26.2. □

The bound for $A_\lambda(r)$ is unbounded with respect to r for $\lambda = 1$. while it is bounded for every $\lambda > 1$. In fact, the former becomes

$$A_\lambda(1-0) = (1-\alpha)^2\, 4\pi\left(\frac{a}{2}\right)^\lambda \zeta\left(\lambda;\frac{a-1}{2}\right),$$

ζ denoting the generalized Riemann zeta function.

§ 27. Range of functions in Carathéodory class

In the integral operator

$$(27.1) \qquad \mathcal{L} f(z) = \int_I \frac{f(zt)}{t} \, d\sigma(t),$$

a particular case generated by $\sigma(t) = t$ is distinguished. Then the operator

$$(27.2) \qquad \{\mathcal{L}^{\lambda}\}: \quad \mathcal{L}^{\lambda} f(z) = \int_I \frac{f(zt)}{t} \, d\sigma_{\lambda}(t)$$

is explicitly determined by

$$(27.3) \qquad \sigma_{\lambda}(t) = \frac{1}{\Gamma(\lambda)} \int_0^t \left(\log \frac{1}{t}\right)^{\lambda-1} dt.$$

and the operator $\mathcal{L}^{\lambda}$ reduces to the fractional integration of order λ with respect to $\log z$, i. e.,

$$\mathcal{L}^{\lambda} f(z) = \int_{\infty}^{\log z} f(e^{\omega})(\log z - \omega)^{\lambda-1} d\omega,$$

the path of integration being taken along the half straight line parallel to the real axis which is contained in the half-plane.

Now, let $P(\alpha)$ with $\alpha < 1$ denote the Carathéodory class of order α which consists of analytic functions p holomorphic in E and satisfying $p(0) = 1$ and $\operatorname{Re} p(z) > \alpha$ in E. It is readily seen that $f(z)/z \in P(\alpha)$ implies

$$\frac{f_\lambda(z)}{z} \in P(\alpha);$$

here and also in the following lines we write $f_\lambda = \mathcal{L}^\lambda f$ for the sake of brevity.

We consider the subclass $\mathcal{F}(\alpha)$ of $\mathcal{F}$ which consists of functions f satisfying $f(z)/z \in P(\alpha)$. We begin with a general theorem on the range of $f_\lambda(z)/z$ for $\{|z| \leqq r\}$. For the discussions in the following lines, cf. Komatu [10, 12, 15].

THEOREM 27.1. *Any function* $f \in \mathcal{F}(\alpha)$ *satisfies*

$$(27.4) \qquad \left| \frac{f_\lambda(z)}{z} - \frac{\varphi_\lambda(r;\alpha)}{r} \right| \leqq \frac{\psi_\lambda(r;\alpha)}{r} - 1$$

for $\{|z| \leqq r < 1\}$, *where* φ *and* ψ *are elementary functions in* $\mathcal{F}$ *defined by*

$$\frac{\chi(z;\alpha)}{z} = (1-\alpha)\frac{1+z}{1-z} + \alpha,$$

$$\frac{\varphi(z\,;\,\alpha)}{z} = \frac{\chi(z^2;\,\alpha)}{z^2} = (1-\alpha)\frac{1+z^2}{1-z^2} + \alpha,$$

$$\frac{\psi(z\,;\,\alpha)}{z} = 1 + (1-\alpha)z + \frac{\chi(z^2;\,\alpha)}{z}$$

$$= 1 + 2(1-\alpha)\frac{z}{1-z^2}.$$

The extremal functions for the estimation are of the form $f(z) = \bar{\varepsilon}\chi(\varepsilon z\,;\,\alpha)$ *with* $|\varepsilon| = 1$, *unless* σ *coincides with the point measure concentrated at* 0. *Further, the range of* $f_\lambda(z)/z$ *for* $\{|z| \leqq r\}$ *induced from any function* f *of this form is just the closed circle expressed by the estimation.*

Proof. Since $f \in \mathcal{J}(\alpha)$ implies $(f - \alpha z)/(1-\alpha) \in \mathcal{J}(0)$, we get in view of Herglotz representation on $P(0)$ the expression

$$\frac{f(z)}{z} = (1-\alpha)\int_{-\pi}^{\pi} \frac{e^{i\theta}+z}{e^{i\theta}-z}\,d\tau(\theta) + \alpha$$

$$= \int_{-\pi}^{\pi} \frac{\chi(e^{-i\theta}z\,;\,\alpha)}{e^{-i\theta}z}\,d\tau(\theta),$$

where τ is a probability measure supported on the interval $(-\pi, \pi]$. Now, the range of $\chi(z\,;\,0)/z \in P(0)$ for $\{|z| \leqq r\}$ is contained in the closed circle with the segment

$$\left[\frac{\chi(-r;0)}{-r}, \frac{\chi(r;0)}{r} \right]$$

as a diameter. Hence, the range of $\chi(z;\alpha)/z$ is contained in the closed circle with the segment

$$\left[\frac{\chi(-r;\alpha)}{-r}, \frac{\chi(r;\alpha)}{r} \right]$$

as a diameter, of which the center and the radius are given by

$$\frac{1}{2}\left(\frac{\chi(r;\alpha)}{r} + \frac{\chi(-r;\alpha)}{-r} \right)$$

$$= (1-\alpha)\frac{1+r^2}{1-r^2} + \alpha = \frac{\varphi(r;\alpha)}{r}$$

and

$$\frac{1}{2}\left(\frac{\chi(r;\alpha)}{r} - \frac{\chi(-r;\alpha)}{-r} \right)$$

$$= 2(1-\alpha)\frac{r}{1-r^2} = \frac{\psi(r;\alpha)}{r} - 1,$$

respectively. Consequently, we have

$$\left| \frac{\chi(z;\alpha)}{z} - \frac{\varphi(r;\alpha)}{r} \right| \leqq \frac{\psi(r;\alpha)}{r} - 1$$

for $|z| \leqq r$. On the other hand, by taking the definition of $\mathcal{L}^{\lambda}$ into account, we obtain

$$\frac{f_\lambda(z)}{z} - \frac{\varphi_\lambda(r;\alpha)}{r}$$

$$= \int_I \left(\int_{-\pi}^{\pi} \frac{\chi(e^{-i\theta}zt;\alpha)}{e^{-i\theta}zt}\, d\tau(\theta) - \frac{\varphi(rt;\alpha)}{rt} \right) d\sigma_\lambda(t)$$

$$= \int_I \left(\int_{-\pi}^{\pi} \left(\frac{\chi(e^{-i\theta}zt;\alpha)}{e^{-i\theta}zt} - \frac{\varphi(rt;\alpha)}{rt} \right) d\tau(\theta) \right) d\sigma_\lambda(t).$$

Thus, by remembering the above inequality, we get the desired inequality

$$\left| \frac{f_\lambda(z)}{z} - \frac{\varphi_\lambda(r;\alpha)}{r} \right|$$

$$\leqq \int_I \left(\int_{-\pi}^{\pi} \left(\frac{\psi(rt;\alpha)}{rt} - 1 \right) d\tau(\theta) \right) d\sigma_\lambda(t)$$

$$= \int_I \left(\frac{\psi(rt;\alpha)}{rt} - 1 \right) d\sigma_\lambda(t)$$

$$= \frac{\psi_\lambda(r;\alpha)}{r} - 1.$$

Concerning the extremal functions, it is readily seen that the equality sign at a point on $\{|z| \leqq r\}$ and necessarily on $\{|z| = r\}$ in the estimation holds if and only if τ is the point measure concentrated at a single point θ, and hence f reduces to $f(z) = \bar{\varepsilon}\chi(\varepsilon z;\alpha)$ with $\varepsilon = e^{-i\theta}$. □

In a previous section, we have introduced the family of operators $\{\mathcal{L}(a)^{\lambda}\}_{\lambda=0}^{\infty}$ generated by the measure $\sigma(t\,;\,a) = t^{a}$ depending on a parameter $a > 0$. It can be explicitly expressed by the representation

$$\mathcal{L}(a)^{\lambda} f(z) = \frac{a^{\lambda}}{\Gamma(\lambda)} \int_I f(zt)\, t^{a-2} \left(\log \frac{1}{t}\right)^{\lambda-1} dt.$$

In the following lines, we shall specialize the result stated in the Theorem 27.1:

THEOREM 27.2. *In the case generated by the measure* $\sigma(t\,;\,a)$, *any function* $f \in \mathcal{F}$ *satisfies*

$$(27.5)\qquad \left| \frac{\mathcal{L}(a)^{\lambda} f(z)}{z} - \frac{\mathcal{L}(a)^{\lambda}\varphi(r\,;\,\alpha)}{r} \right| \leqq \frac{\mathcal{L}(a)^{\lambda}\psi(r\,;\,\alpha)}{r} - 1$$

for $|z| \leqq r < 1$, *where the quantities involved refer to those induced from* φ *and* ψ *in Theorem* 27.1 *and are given in series form*

$$\frac{\mathcal{L}(a)^{\lambda}\varphi(r\,;\,\alpha)}{r} = 1 + 2(1-\alpha) \sum_{n=1}^{\infty} \left(\frac{a}{2n+a}\right)^{\lambda} r^{2n},$$

$$\frac{\mathcal{L}(a)^{\lambda}\psi(r\,;\,\alpha)}{r} = 1 + 2(1-\alpha) \sum_{n=1}^{\infty} \left(\frac{a}{2n+a-1}\right)^{\lambda} r^{2n-1}.$$

The extremal functions for the estimation are of the form $f(z) = \bar{\varepsilon}\chi(\varepsilon z;\alpha)$ *with* $|\varepsilon| = 1$ *and the same* χ *as in Theorem* 27.1.

Proof. We only have to verify the series expansions for $\mathcal{L}(a)^{\lambda}\varphi$ and $\mathcal{L}(a)^{\lambda}\psi$. By making use of the series expansions for φ and ψ, we get

$$\frac{\mathcal{L}(a)^{\lambda}\varphi(z;\alpha)}{z}$$

$$= \frac{a^{\lambda}}{\Gamma(\alpha)}\int_I \left(1 + 2(1-\alpha)\sum_{n=1}^{\infty} z^{2n} t^{2n}\right) t^{a-1}\left(\log\frac{1}{t}\right)^{\lambda-1} dt,$$

$$\mathcal{L}(a)^{\lambda}\psi(z;\alpha)$$

$$= \frac{a^{\lambda}}{\Gamma(\lambda)}\int_I \Bigg(1$$

$$+ 2(1-\alpha)\sum_{n=1}^{\infty} z^{2n-1} t^{2n-1}\Bigg) t^{a-1}\left(\log\frac{1}{t}\right)^{\lambda-1} dt.$$

Then, by integrating termwise and taking into account a familiar formula

$$\int_I t^{\nu-1}\left(\log\frac{1}{t}\right)^{\lambda-1} dt = \frac{\Gamma(\lambda)}{\nu^{\lambda}},$$

we obtain, after putting $z = r$, the desired expansions. In assertion on extremal function, the extreme case does not appear. □

Finally, we supplement a theorem on the range in the whole disk:

THEOREM 27.3. *In the case generated by* $\sigma(t\,;\,a)$, *the range of values of* $(\mathcal{L}(a)^{\lambda}f(z))/z$ *with* $\lambda > 1$ *in* E *for any* $f \in \mathcal{J}$ *is contained within the circular disk with the segment*

$$\left(1 - 2(1-\alpha)\left(\frac{a}{2}\right)^{\lambda}\left(\zeta\left(\lambda;\frac{a+1}{2}\right) - \zeta\left(\lambda;\frac{a+2}{2}\right)\right),\right.$$

$$\left.1 + 2(1-\alpha)\left(\frac{a}{2}\right)^{\lambda}\left(\zeta\left(\lambda;\frac{a+1}{2}\right) - \zeta\left(\lambda;\frac{a+2}{2}\right)\right)\right)$$

as a diameter, i. e., the estimation

$$\left|\frac{\mathcal{L}(a)^{\lambda}f(z)}{z} - \left(1 + 2(1-\alpha)\left(\frac{a}{2}\right)^{\lambda}\zeta\left(\lambda;\frac{a+2}{2}\right)\right)\right|$$

$$< 2(1-\alpha)\left(\frac{a}{2}\right)^{\lambda}\zeta\left(\lambda;\frac{a+1}{2}\right)$$

holds for any $z \in E$, *where* ζ *denotes the generalized Riemann zeta function*.

Proof. The range-circle for $(\mathcal{L}(a)^{\lambda}f(z)/z$ $(|z| \leq r)$ in Theorem 27.1 swells as point set together with $r \in [0, 1)$. If $\lambda > 1$, its center and radius tend to

$$1 + 2(1 - \alpha) \sum_{n=1}^{\infty} \left(\frac{a}{2n + a} \right)^{\lambda} = 2(1 - \alpha) \left(\frac{a}{2} \right)^{\lambda} \zeta\left(\lambda; \frac{a + 2}{2}\right)$$

and

$$2(1 - \alpha) \sum_{n=1}^{\infty} \left(\frac{a}{2n + a - 1} \right)^{\lambda} = 2(1 - \alpha) \left(\frac{a}{2} \right)^{\lambda} \zeta\left(\lambda; \frac{a + 1}{2}\right).$$

respectively, as $r \to 1 - 0$. Hence the result follows. □

Now, when $0 < \lambda \leqq 1$, both center and radius diverge to positive infinity as $r \to 1 - 0$. However, the left end-point of the diameter lying on the real axis remains finite; in fact, it lies always on the left of the point α. Moreover, the range in this case can be determined explicitly as shown in the following theorem:

THEOREM 27.4. *In case of* $0 < \lambda < 1$, *instead of* $\lambda > 1$, *the range observed in the previous Theorem* 26.3 *is contained in the half - plane lying on the right side of the point*

$$u(\lambda; \alpha) := 1 - 2(1 - \alpha) \sum_{n=1}^{\infty} (-1)^{n-1} \left(\frac{a}{n + a} \right)^{\lambda},$$

i. e., the estimation

$$\mathrm{Re} \frac{\mathcal{L}(a)^{\lambda} f(z)}{z} > u(\lambda; \alpha)$$

holds for any $z \in E$.

Proof. In view of the notice mentioned just above, we have only to verify that the range-circle stated in Theorem 26.2 tends to the half-plane as $r \to 1-0$. Now, the left endpoint of the range-circle for $|z| \leqq r$ lies at

$$\frac{\mathcal{L}(a)^{\lambda}\varphi(r;\alpha)}{r} - \left(\frac{\mathcal{L}(a)^{\lambda}\psi(r;\alpha)}{r} - 1\right)$$

$$= 1 - 2(1-\alpha)\sum_{n=1}^{\infty}(-1)^{n-1}\left(\frac{a}{n+a}\right)^{\lambda} r^{n}.$$

By taking Abel's continuity theorem into account, we obtain

$$\lim_{r\to 1-0}\left(\frac{\mathcal{L}(a)^{\lambda}\varphi(r;\alpha)}{r} - \left(\frac{\mathcal{L}(a)^{\lambda}\psi(r;\alpha)}{r} - 1\right)\right)$$

$$= 1 - 2(1-\alpha)\sum_{n=1}^{\infty}(-1)^{n-1}\left(\frac{a}{n+a}\right)^{\lambda} = u(\lambda;\alpha),$$

as desired. □

Finally, we observe the case where a is a positive integer k. Then, we have

$$u(\lambda;\alpha)$$

$$= 1 - 2(1-\alpha)(-1)^{k} k^{\lambda}\left((1-2^{1-\lambda})\zeta(\lambda) - \sum_{n=1}^{k}\frac{(-1)^{n-1}}{n^{\lambda}}\right),$$

where ζ denotes the ordinary Riemann zeta function analytically prolonged. When $\lambda = 1$, this becomes

$$u(1;\alpha) = 1 + 2(1-\alpha)(-1)^{k-1} k\left(\log 2 - \sum_{n=1}^{k} \frac{(-1)^{n-1}}{n}\right).$$

On the other hand, when $\lambda = 0$, we have

$$u(0;\alpha) = \alpha,$$

while the value of the derivative $(\partial/\partial\lambda)u(\lambda;\alpha)$ at $\lambda = 0$ is equal to

$$\left[\frac{\partial}{\partial\lambda} u(\lambda;\alpha)\right]^{\lambda=+0}$$

$$= -(1-\alpha)\log k$$

$$+ 2(1-\alpha)(-1)^{k-1}\left(\frac{1}{2}\log\frac{\pi}{2} + \sum_{n=2}^{k}(-1)^{n-1}\log n\right)$$

$$= (1-\alpha)\left(2(-1)^{k-1}\log\frac{k!\,(\pi/2)^{1/2}}{([k/2]!\,2^{[k/2]})^2} - \log k\right),$$

which coincides with $(1-\alpha)\Psi(k)$ where Ψ is given by

$$\Psi(k) = \left[\frac{\partial\,\Phi(\delta, k)}{\partial\delta}\right]^{\delta=+0},$$

$$\Phi(\delta, k) = \int_I \frac{1-t}{1+t}\, d\sigma_\delta(t;k).$$

In particular, if $k = 1$,

$$u(\lambda;\ \alpha) = 2(1-\alpha)(1-2^{1-\lambda})\zeta(\lambda) - (1-2\alpha),$$

$$u(1;\ \alpha) = 2(1-\alpha)\log 2 - (1-2\alpha)$$

and the value of its derivative at $\lambda = 0$ becomes

$$\left[\frac{\partial}{\partial\lambda} u(\lambda;\ \alpha)\right]^{\lambda=0} = (1-\alpha)\log\frac{\pi}{2}.$$

REMARK. It is noted that the way followed here is believed as a model showing how to deal with similar problems concerning linear functionals of f in several subclasses of $\mathcal{J}$.

§ 28. A general operator on Carathéodory class

The Carathéodoty class $P(0)$, which consists of functions ϕ holomorphic and satisfying $\mathrm{Re}\ \phi > 0$ in E and normalized by $\phi(0) = 1$ at the origin, has been dealt with hitherto very often. It admits the representation of Herglotz [1]:

$$\phi(z) = \int_{-\pi}^{\pi} \frac{e^{i\varphi} + z}{e^{i\varphi} - z}\, d\rho(\varphi),$$

where ρ is a probability measure defined on $(-\pi, \pi]$:

$$d\rho(\varphi) \geqq 0 \quad (-\pi, \pi], \qquad \int_{-\pi}^{\pi} d\rho(\omega) = 1.$$

This representation will be frequently referred to also in the subsequent arguments.

In the present section, we consider a general linear operator $\mathcal{L}[\ ]$ which has $\mathcal{P}(0)$ as its domain of argument functions and the class of analytic functions single-valued about the origin as its range. Suppose that the operator is homogeneous of degree zero, i. e., for any constant c, the function $\mathcal{L}[\phi(z)]$ coincides after substitution $z \mid cz$ with $\mathcal{L}[\varphi(cz)]$; in particular,

$$\mathcal{L}_z\left[\frac{e^{i\varphi}+z}{e^{i\varphi}-z}\right] = \mathcal{L}_z\left[\frac{1+e^{-i\varphi}z}{1-e^{-i\varphi}z}\right] = \mathcal{L}_\zeta\left[\frac{1+\zeta}{1-\zeta}\right]^{\zeta=e^{-i\varphi}z}.$$

Together with an operator, we observe an increasing function F which is convex in $[0, +\infty)$; concerning the present section, cf. Komatu [1].

Our first problem to be discussed is to obtain the precise estimate for the functional defined by

$$\int_{-\pi}^{\pi} F(|\mathcal{L}[\phi(re^{i\theta})]|)\,d\theta \quad (0 < r < 1)$$

within the class $P(0)$ in terms of a definite function of r and further to determine the extremal functions for the estimation.

Problems in this category have been hitherto dealt with, for instance, in Lemma 25.1. It will be shown that the functional under consideration is majorized by its value attained by substituting $(1 + \varepsilon z)/(1 - \varepsilon z)$ with $|\varepsilon| = 1$ for $\phi(z)$, i. e., the linear function of this form possesses always the extremal character. However, other extremal functions can appear for certain operator $\mathcal{L}$. Accordingly, the circumstances are not so simple if we attempt to determine the whole of extremal functions. So, it will be convenient to establish some preparatory lemmata which serve for this purpose.

We now begin with a Lemma on an elementary mapping which will play in the principal lemma an auxiliary role of cancelling the zero-points or poles of a function without altering its argument.

LEMMA 28.1. *Let c be a point belonging to the closed circular disk $\{|z| \leq r\}$. Then, the function $\omega(z, c)$ which maps $\{|z| < r\}$ onto the whole plane cut along a slit lying on the positive real axis, and normalized by $\omega(0, c) = \infty$, $\omega(c, c)$ and $|\mathrm{Res}\,(0;\omega)| = r^2$ is uniquely determined, and its explicit expression is given by*

$$\omega(z, c) = e^{i \arg c}(c - z)\left(1 - \frac{r^2}{\bar{c} z}\right).$$

Let further n be any positive integer. Then, the function defined by

$$\Omega(z, c) = \prod_{k=0}^{n-1} \omega(z, ce^{2k\pi i/n})$$

is a single - valued function of z^n holomorphic for $z \neq 0$

but not of any higher power of z.

Proof. The uniqueness assertion of the mapping function is readily verified, for instance, by means of a theorem on radial slit mapping. That it is expressed as written in the Lemma may be shown directly. In fact, we have

$$\omega(re^{i\theta}, c) = \frac{1}{|c|}|c - re^{i\theta}|^2,$$

so that the image of $\{|z| = r\}$ by $\omega(z, c)$ is a segment on the positive real axis. Since ω is holomorphic except a simple pole at the origin, it is surely univalent in $\{|z| < r\}$. It is evident that the normalization conditions are also satisfied.

Next, it is readily seen from its explicit expression that ω satisfies for any real φ the relation

$$\omega(ze^{i\varphi}, c) = \omega(z, ce^{-i\varphi})$$

which implies

$$\Omega(ze^{2\pi i/n}, c) = \prod_{k=0}^{n-1} \omega(ze^{2\pi i/n}, ce^{2k\pi i/n})$$

$$= \prod_{k=0}^{n-1} \omega(z, ce^{2(k-1)\pi i/n}) = \Omega(z, c).$$

Thus, Ω being invariant under the substitution $z \mapsto ze^{2\pi i/n}$, it is a single-valued function of z^n. Since it has a pole of order n at the origin, it cannot be a single-valued function of a higher power of z . □

Now, we mention the principal Lemma which is fundamental for our subsequent discussions:

LEMMA 28.2. *Let* Ψ *be an analytic function meromorphic in* $\{|z| < r\}$ *which is non-constant and holomorphic along* $\{|z| = r\}$. *Let* n *denote the greatest integer such that* $\Psi(z)$ *is a single-valued function of* z^n. *Further, let* ρ *be a realvalued function defined for* $(-\pi, \pi]$ *which is increasing and of bounded variation with positive variation. Then, in order that the relation*

$$\left| \int_{-\pi}^{\pi} \Psi(re^{i(\theta-\varphi)})\, d\rho(\varphi) \right| = \int_{-\pi}^{\pi} |\Psi(re^{i(\theta-\varphi)})|\, d\rho(\varphi)$$

holds identically with respect to θ *throughout* $(-\pi, \pi]$, *it is necessary and sufficient that* ρ *remains unchanged except possibly at* mn *jump points which are distributed equidistantly in* $(-\pi, \pi]$. *Here,* m *is a positive integer defined as follows : Let the sets of zero-points and of poles of* Ψ *contained in* $\{0 < |z| \leq r\}$ *which are irreducible with respect to their arguments taken by* mod $2\pi/n$ *be* $\{a_\lambda\}_{\lambda=1}^{\alpha}$ *and* $\{b_\mu\}_{\mu=1}^{\beta}$, *respectively, which are counted according to respective multiplicities and let* N *be the greatest integer such that the function defined by*

$$Y(z) = \Psi(z) \prod_{\mu=1}^{\beta} \Omega(z, b_\mu) \Big/ \prod_{\lambda=1}^{\alpha} \Omega(z, a_\lambda)$$

is a single - valued function of z^N, Ω *being the function introduced in the Lemma* 28.1. *Then we put* $N = mn$.

Proof. The sufficiency assertion is readily verified. In fact, let ρ remain unchanged except at mn jumps

$$\varphi_0 + \varphi_{jk} = \varphi_0 + 2((j-1)n + k)\pi/mn$$

$$(0 \leqq k \leqq n-1,\ 1 \leqq j \leqq m)$$

with the heights $\rho_{jk} \geqq 0$, respectively. Since Y is single-valued in z^{mn}, we then have

$$\Psi\left(re^{i(\theta-\varphi_0-\varphi_{jk})}\right)$$

$$= Y(re^{i(\theta-\varphi_0)}) \prod_{\lambda=1}^{\alpha} \Omega\left(re^{i(\theta-\varphi_0-\varphi_{jk})}, a_\lambda\right)$$

$$\Big/ \prod_{\mu=1}^{\beta} \Omega\left(re^{i(\theta-\varphi_0-\varphi_{jk})}, b_\mu\right),$$

so that, by virtue of $\arg \Omega(z, c) = 0$ valid along $\{|z| = r\}$, there follows

$$\left| \int_{-\pi}^{\pi} \Psi(re^{i(\theta-\varphi)})\, d\rho(\varphi) \right|$$

$$= |Y(re^{i(\theta-\varphi_0)})| \sum_{j=1}^{m} \sum_{k=0}^{n-1} \rho_{jk} \prod_{\lambda=1}^{\alpha} \Omega\left(re^{i(\theta-\varphi_0-\varphi_{jk})}, a_\lambda\right)$$

$$\Big/ \prod_{\mu=1}^{\beta} \Omega\left(re^{i(\theta-\varphi_0-\varphi_{jk})}, b_\mu\right)$$

$$= \int_{-\pi}^{\pi} |\Psi(re^{i(\theta-\varphi)})| \, d\rho(\varphi).$$

The necessity proof proceeds as follows. If the relation under consideration holds identically with respect to θ, the quantity $\arg \Psi(ze^{-i\varphi})$ with any fixed $z = re^{i\theta}$ must have the same value depending only on θ for every value of φ with $d\rho(\varphi) > 0$ provided $\Psi(ze^{-i\varphi})$ does not vanish. Hence, $\Omega(z, c)$ being remain to be real and positive along $\{|z| = r\}$, $\arg Y(ze^{-i\varphi})$ has also the same property. Let φ_0 and φ_1 be any value of φ with $d\rho(\varphi) > 0$, and we observe the function X defined by

$$X(z) = \frac{Y(ze^{-i\varphi_0})}{Y(ze^{-i\varphi_1})}.$$

Since Y is holomorphic and non-vanishing throughout $\{0 < |z| \le r\}$, X is holomorphic for $\{|z| \le r\}$ even at the origin. Further, it remains real and positive along $\{|z| = r\}$. In general, any bounded set lying entirely on the real axis cannot be the boundary of the image of $\{|z| \le r\}$ by a (not necedssarily univalent) mapping holomorphic there unless it degenerates to a single point. Consequently, X must be a constant which is real and positive. Let the Laurent expansion of Y be

$$Y(z) = \sum c_\nu z^{n_\nu}$$

where $\{n_\nu\}$ is a strictly increasing (finite or infinite) sequence of integers for which the sequence of corresponding coefficients $\{c_\nu\}$ does not involve zero. We then get

$$Y(ze^{-i\varphi_0}) = X(0)\,Y(ze^{-i\varphi_1})$$

and obtain further, by comparing the coefficients of z^{n_ν},

$$e^{-in_\nu\varphi_0} = X(0)\,e^{-in_\nu\varphi_1}$$

for any ν. Since $X(0)$ is real and positive, we get $X(0) = 1$, whence follows

$$n_\nu(\varphi_1 - \varphi_0) \equiv 0 \pmod{2\pi}$$

for any ν. Now, since $N = mn$ is equal to the greatest common measure of the set $\{n_\nu\}$, there exist two members of the set, n_γ and n_δ say, such that N is just the greatest common measure of them. Hence, for some integers u and v, we have $un_\gamma - vn_\delta = N$. Consequently, the above relations applied to $n_\nu = n_\gamma$ and to $n_\nu = n_\delta$ imply

$$N(\varphi_1 - \varphi_0) \equiv 0 \pmod{2\pi}.$$

Thus, φ_0 and φ_1 being arbitrary pair where $d\rho(\varphi)$ does not vanish, it has been verified that ρ must possess the property asserted. □

For subsequent purpose, it will become necessary to know the actual values of the integers n and m defined in the Lemma 28.2. Accordingly, as a supplement of this Lemma, we pick out particular cases for which we formulate the following Lemma:

LEMMA 28.3. *In Lemma* 28.2, *if* $\{a_\lambda\}$ *and* $\{b_\mu\}$ *are, in particular, both vacuous, namely, if* $\Psi(z)$ *is hlomorphic and non - vanishing throughout* $\{0 < |z| \leq r\}$, *then we have* $m = 1$. *Further, if* $\Psi'(0) \neq 0$, *then we have* $n = 1$.

It may be noted by the way that, if Ψ has zero-points or poles, then m may be actually greater than unity. It will be illustrated by an example. Let $0 < |a| < r$ and put, as before,

$$\Omega(z, a) = \prod_{k=1}^{n-1} \omega(z, ae^{2k\pi i/n}).$$

Then, this is a single-valued function of z^n but not of any higher power of z. Hence, the function defined by

$$\Psi(z) = z^{mn}\Omega(z, a)$$

with any positive integer m is also of the same nature. But, for any ρ with jumps alone at mn values

$$\varphi_{jk} = \frac{2((j-1)n+k)}{mn} \qquad (0 \leq k \leq n-1, \quad 1 \leq j \leq m)$$

with any respective heights ρ_{jk}, we get

$$\left| \int_{-\pi}^{\pi} \psi(re^{i(\theta-\varphi)}) \, d\rho(\varphi) \right|$$

$$= \left| \sum_{j=1}^{m} \sum_{k=0}^{n-1} \psi\left(re^{i(\theta-\varphi_{jk})}\right) \rho_{jk} \right|$$

$$= \sum_{j=1}^{m} \sum_{k=0}^{n-1} r^{mn} \Omega\left(re^{i(\theta-\varphi_{jk})}\right) \rho_{jk}$$

$$= \sum_{j=1}^{m} \sum_{k=0}^{n-1} \left| \psi\left(re^{i(\theta-\varphi_{jk})}\right) \right| \rho_{jk}$$

$$= \int_{-\pi}^{\pi} |\psi(re^{i(\theta-\varphi)})| \, d\rho(\varphi).$$

We are now in position to formulate our main Theorems:

THEOREM 28.1. *Let $\mathcal{L}$ denote a linear operator defined on $\mathcal{F}(0)$ which produces an analytic function meromorphic in $\{|z| < r\,(<1)\}$ and holomorphic along $\{|z| = r\}$, and let be commutable with the integration with respect to the probability measure ρ in the Herglotz representation for ϕ. Let further F be a bounded increasing and convex function de-*

fined for the range of $|\mathcal{L}|$. *Then, for any* $\phi \in \mathcal{J}(0)$, *we have*

$$\int_{-\pi}^{\pi} F(|\mathcal{L}[\phi(re^{i\theta})]|)\, d\theta \leq \int_{-\pi}^{\pi} F(|\mathcal{L}[\phi^*(re^{i\theta})]|)\, d\theta,$$

where $\phi^*(z) = (1+z)/(1-z)$. *The function* $\phi^*(\varepsilon z)$ *with* $|\varepsilon| = 1$ *is always extremal for the estimation.*

Proof. In view of the commutability of $\mathcal{L}$ and the integration with respect to ρ implies

$$|\mathcal{L}[\phi(re^{i\theta})]| = \left|\int_{-\pi}^{\pi} \mathcal{L}[\phi^*(re^{i(\theta-\varphi)})]\, d\rho(\varphi)\right|$$

$$\leq \int_{-\pi}^{\pi} \left|\mathcal{L}[\phi^*(re^{i(\theta-\varphi)})]\right|\, d\rho(\varphi),$$

and the increasing character and the convexity of F imply

$$F(|\mathcal{L}[\phi(re^{i\theta})]|) \leq F\left(\int_{-\pi}^{\pi} |\mathcal{L}[\phi^*(re^{i(\theta-\varphi)})]|\right) d\rho(\varphi)$$

$$\leq \int_{-\pi}^{\pi} F(|\mathcal{L}[\phi^*(re^{i(\theta-\varphi)})]|)\, d\rho(\varphi),$$

since ρ is increasing for $(-\pi, \pi]$ and has the total variation equal to unity. Hence, integrating with resoect to θ, we obtain

$$\int_{-\pi}^{\pi} F(|\mathcal{L}[\phi(re^{i\theta})]|)\, d\theta$$

$$\leq \int_{-\pi}^{\pi} d\theta \int_{-\pi}^{\pi} F(|\mathcal{L}[\phi^*(re^{i(\theta-\varphi)})]|)\, d\rho(\varphi)$$

$$= \int_{-\pi}^{\pi} d\rho(\varphi) \int_{-\pi}^{\pi} F(|\mathcal{L}[\phi^*(re^{i(\theta-\varphi)})]|)\, d\theta$$

$$= \int_{-\pi}^{\pi} F(|\mathcal{L}[\phi^*(re^{i\theta})]|)\, d\theta.$$

For $\phi^*(\varepsilon z)$ with $|\varepsilon| = 1$, it is evident that the equality sign holds surely in the estimation. □

In dealing with extremal functions in the estimarion mentioned in Theorem 27.1, we first consider especially the case where F is linear. Then, in view of the homogeneity of the relation to be considered, we have only to observe the case $F(X) = X$:

THEOREM 28.2. *Under the conditions imposed on* $\mathcal{L}$ *in Theorem 28.1, suppose further that* $\mathcal{L}[\phi^*(z)]$ *is non - constant and let* n *denote the greatest integer such that it is a single - valued function of* z^n. *Then, the relation*

$$\int_{-\pi}^{\pi} |\mathcal{L}[\phi(re^{i\theta})]|\, d\theta = \int_{-\pi}^{\pi} |\mathcal{L}[\phi^*(re^{i\theta})]|\, d\theta$$

holds if and only if ϕ *is of the form*

$$\phi(z) = \sum_{j=1}^{m} \sum_{k=0}^{n} \rho_{jk}\, \phi^*\left(\exp\left(-\frac{2(j-1)n+k}{mn}\pi i\right)\varepsilon z\right)$$

where ε *is a constant with* $|\varepsilon| = 1$ *and* $\{\rho_{jk}\}$ *is a set of real numbers satisfying*

$$\rho_{jk} \geq 0 \quad (0 \leq k \leq n-1,\ 1 \leq j \leq m), \qquad \sum_{j=1}^{m} \sum_{k=0}^{n-1} \rho_{jk} = 1,$$

and m *is a positive integer defined in Lemma* 27.2 *in which* $\Psi(z)$ *is replaced by* $\mathcal{L}[\phi^*(z)]$.

Proof. Based on the proof given above for Theorem 27.1, we see that the extremal character of $\phi(z)$ is characterized in terms of its associated function $\rho(\varphi)$ by the requirement

$$\left| \int_{-\pi}^{\pi} \mathcal{L}[\phi^*(re^{i(\theta-\varphi)})]\, d\rho(\varphi) \right|$$

$$= \int_{-\pi}^{\pi} |\mathcal{L}[\phi^*(re^{i(\theta-\varphi)})]|\, d\rho(\varphi)$$

to be valid identically with respect to θ. We now apply the Lemma 28.2 to the function defined by

$$\Psi(z) = \mathcal{L}\left[\frac{1+z}{1-z}\right],$$

whence follows that for any extremal function ϕ its associated function ρ remains unchanged except possibly at $\varphi_0 + 2((j-1)n+k)\pi/mn$ $(0 \leq k \leq n-1,\ 1 \leq j \leq m)$ with heights $\rho_{jk} \geq 0$, respectively, say; here m, n and $\{\rho_{jk}\}$ satisfy the conditions in the Theorem. Thus, any extremal function must have the form stated in the Theorem with $\varepsilon = e^{-i\varphi_0}$. Conversely, for any function of such a form, the relation under consideration holds good, as seen in the proof of Lemma 28.2.

In case F is non-linear and hence strictly convex, the condition in Theorem 28.2 imposed on extremal function is to be modified by supplementing a further condition:

THEOREM 28.3. *Under the conditions imposed on* $\mathcal{L}$ *and* F *in Theorem* 28.1, *let further the increasing function* F *be strictly convex. Then, the equality sign in the estimation given in Theorem* 28.1 *holds for any fixed* r *if and only if* ϕ *is of the form given in Theorem* 28.2 *and all those members among* m *quantities*

$$\left|\mathcal{L}\left[\phi^*\left(r\exp\left(\frac{i(\theta-\varphi_0-2(j-1)\pi/m}{m}\right)\right]\right| \quad (j = 1, \ldots, m),$$

which correspond to the non-vanishing $\sum_{k=0}^{n-1} \rho_{jk}$, *have the same value for any* θ *throughout* $(-\pi, \pi]$.

Proof. The condition that the equality sign appears in the estimation given in Theorem 28.1 is equivalent to the requirement that the function ρ associated to an extremal function ϕ satisfies, beside the relation

$$\left|\int_{-\pi}^{\pi} \mathcal{L}[\phi^*(re^{i(\theta-\varphi)})]\, d\rho(\varphi)\right|$$

$$= \int_{-\pi}^{\pi} |\mathcal{L}[\phi^*(re^{i(\theta-\varphi)})|\ d\rho(\varphi),$$

a further relation

$$F\left(\int_{-\pi}^{\pi} |\mathcal{L}[\phi^*(re^{i(\theta-\varphi)})]|\ d\rho(\varphi)\right)$$

$$= \int_{-\pi}^{\pi} F(|\mathcal{L}[\phi^*(re^{i(\theta-\varphi)})]|)\ d\rho(\varphi)$$

also identically with respect to θ. The first relation implies the consequence stated in Theorem 28.2. Now, for any function of the form written in Theorem 28.2, we get

$$F\left(\int_{-\pi}^{\pi} |\mathcal{L}[\phi^*(re^{i(\theta-\varphi)})|\ d\rho(\varphi)\right)$$

$$= F\left(\sum_{j=1}^{m}\sum_{k=0}^{n-1} \rho_{jk} \left|\mathcal{L}\left[\phi^*\left(r\exp\left(i\left(\theta-\varphi_0-\frac{2(j-1)\pi}{m}\right)\right)\right)\right]\right|\right)$$

and

$$\int_{-\pi}^{\pi} F(|\mathcal{L}[\phi^*(re^{i(\theta-\varphi)})]|)\ d\rho(\varphi)$$

$$= \sum_{j=1}^{m}\sum_{k=0}^{n-1} \rho_{jk}\, F\left(\left|\mathcal{L}\left[\phi^*\left(r\exp\left(i\left(\theta-\varphi_0-\frac{2(j-1)\pi}{m}\right)\right)\right)\right]\right|\right),$$

since $\mathcal{L}[\phi^*(z)]$ is a single-valued function of z^n. F being supposed strictly convex, the right-hand members of the last two equations are equal for any r if and only if the last-mentioned condition of the theorem is satisfied. □

Though the following result is an immediate consequence of the Corollary 28.1 combined with Lemma 27.3, we write down it here for its frequent use in the subsequent discussions:

THEOREM 28.4. *Under the conditions imposed on* $\mathcal{L}$ *and* F *in Theorem* 28.1, *let* $\mathcal{L}[\phi^*]$ *be non-constant, holomorphic and non-vanishing for* $0 < |z| \leqq r$ *and* F *be strictly increasing. Then, the equality sign in the estimation given in Theorem* 28.1 *holds for any fixed* r *if and only if* ϕ *is of the form given in the Corollary* 28.1. *If, moreover,* $\mathcal{L}[\phi^*]$ *satisfies an additional condition that its derivative does not vanish at the origin, then the extremal function must be of the form* ϕ^* *with* $|\varepsilon| = 1$.

In Theorems 28.2, 28.3 and 28.4, it has been supposed that $\mathcal{L}[\phi^*]$ does not reduce to a constant. However, it is to be noticed that, if it reduces to a constant, the operator $\mathcal{L}$ degenerates identically to a constant, i. e., its range consists of a single constant function.

Here we supplement a remark on the Corollary 28.1. We now suppose for a while that the operator $\mathcal{L}$ is applicable termwise to the Taylor series of argument function of which the basis $\{z^n\}_{\nu=0}^{\infty}$ does not belong to $P(0)$ except the first term. Then, we get, in particular,

$$\mathcal{L}[\phi^*(z)] = \mathcal{L}[1] + 2\sum_{\nu=1}^{\infty} \mathcal{L}[z^{\nu}].$$

Consequently, the condition that this function is single-valued with respect to z^n is equivalent to the system of conditions $\mathcal{L}[z^\nu] = 0$ for $\nu \not\equiv 0 \pmod{n}$, whence follows, in particular, $\mathcal{L}[\phi^*(z)]$ is then a single-valued function of z^n. Thus, if we put

$$\mathcal{L}[\phi(z)] = \not\phi(z^n)$$

and

$$\mathcal{L}[\phi^*(z)] = \not\phi^*(z^n) = \mathcal{L}[\phi^*(z^n)],$$

the inequality mentioned in Theorem 28.1 becomes

$$\int_{-\pi}^{\pi} F(|\not\phi(r^n e^{in\theta})|)\,d\theta \leq \int_{-\pi}^{\pi} F(|\not\phi^*(r^n e^{in\theta})|)\,d\theta$$

which is equivalent to

$$\int_{-\pi}^{\pi} F(|\not\phi(te^{i\sigma})|)\,d\sigma \leq \int_{-\pi}^{\pi} F(|\not\phi^*(te^{i\sigma})|)\,d\sigma, \quad t = r^n.$$

The last relation coincides formally with that given in Theorem 28.1 with t instead of r and applied to the operator transforming ϕ into $\not\phi$. Corollary 28.1 then asserts that for such an operator the equality sign in the last estimation is realized not only by a function of the form

$$\phi^*(\eta z^n) \quad \text{with} \quad |\eta| = 1$$

but also by any function of the form mentioned there.

In Theorem 28.1, several sorts of choice are possible for $\mathcal{L}$ as well as F. Thus, for instance, we may take as $\mathcal{L}$ any operator which transforms $\phi(z)$ into a linear combination of $\phi^{(\mu)}(\gamma_j z)$ $(\mu = 0, 1, \ldots, M;\ j = 1, \ldots, J)$, γ_j being arbitrary constants with $|\gamma_j| < 1/r$. Among such operators, we choose here as illustrating examples two special ones and mention the following Theorem which involves in itself the Theorem of Rogosinski [1] as a particular case:

THEOREM 28.5. (i) *Let n be a positive integer, ν a non-negative integer and $p \geqq 1$ a real number. Then, for any $\phi \in P(0)$, we have*

$$\frac{1}{n}\int_{-\pi}^{\pi}\left|\sum_{k=0}^{n-1} e^{-2\nu k\pi i/n}\phi^{(\nu)}(re^{i(\theta-2k\pi/n)})\right|^{p} d\theta$$

$$\leqq \int_{-\pi}^{\pi}\left|\frac{\partial^{\nu}}{\partial r^{\nu}}\frac{1+r^{n}e^{i\theta}}{1-r^{n}e^{i\theta}}\right| d\theta \qquad (r \in (0, 1)).$$

The equality sign holds for any fixed r if and only if ϕ is of the form

$$\phi(z) = \sum_{k=0}^{n-1}\rho_k\frac{e^{2k\pi i/n}+\varepsilon z}{e^{2k\pi i/n}-\varepsilon z},$$

where ε and $\{\rho_k\}_{k=0}^{n-1}$ are defined as in the Corollary 27.1.

(ii) *Let n be an even positive integer, ν a non-negative integer and $p \geqq 1$ a real number. Then, for any $\phi \in P(0)$, we have*

$$\frac{1}{n}\int_{-\pi}^{\pi} \left| \sum_{k=0}^{n-1} (-1)^k e^{-2\nu k\pi i/n} \phi^{(\nu)}(re^{i(\theta-2k\pi/n)}) \right|^p d\theta$$

$$\leqq \int_{-\pi}^{\pi} \left| \frac{\partial^\nu}{\partial r^\nu} \frac{2r^{n/2} e^{i\theta}}{1 - r^n e^{2i\theta}} \right|^p d\theta \qquad (r \in (0, 1)).$$

The equality holds for any fixed r if and only if ϕ is of the form

$$\phi(z) = \sum_{\kappa=0}^{n/2-1} \rho_\kappa \frac{e^{4\kappa\pi i/n} + \varepsilon z}{e^{4\kappa\pi i/n} - \varepsilon z}$$

with

$$|\varepsilon| = 1, \quad \rho_\kappa \geqq 0 \ \left(0 \leqq \kappa \leqq \frac{n}{2} - 1\right), \qquad \sum_{\kappa=0}^{n/2-1} \rho_\kappa = 1.$$

Proof. (i) Applying Theorem 28.1 and 27.4 to the pair

$$\mathcal{L}[\phi(z)] = z^\nu \frac{d^\nu}{dz^\nu} \frac{1}{n} \sum_{k=0}^{n-1} \phi(ze^{-2k\pi i/n}),$$

$$F(X) = X^p,$$

the desired result follows readily after reducing both members by $r^{\nu p}$. In fact, we only have to recll the identity

$$\frac{1}{n}\sum_{k=0}^{n-1}\frac{1+ze^{-2k\pi i/n}}{1-ze^{-2k\pi i/n}}=\frac{1}{n}\sum_{k=0}^{n-1}\left(1+2\sum_{h=1}^{\infty}z^{h}e^{-2hk\pi i/n}\right)$$

$$=1+2\sum_{\mu=1}^{\infty}z^{\mu n}=\frac{1+z^{n}}{1-z^{n}}$$

in which the last nember shows, in particular, that $\chi[(1+z)/(1-z)]$ is single-valued in z^n but not in any higher power of z, and then to transform the estimate by changing the integration variable θ into θ/n, whence follows

$$\int_{-\pi}^{\pi}\left|\frac{d^{\nu}}{d(re^{i\theta})^{\nu}}\frac{1+(re^{i\theta})^{n}}{1-(re^{i\theta})^{n}}\right|^{p}d\theta$$

$$=\int_{-\pi}^{\pi}\left|\frac{\partial^{\nu}}{\partial r^{\nu}}\frac{1+r^{n}e^{i\theta}}{1-r^{n}e^{i\theta}}\right|^{p}d\theta.$$

(ii) We can proceed similarly as above. We now take

$$\mathcal{L}[\phi(z)]=z^{\nu}\frac{d^{\nu}}{dz^{\nu}}\frac{1}{n}\sum_{k=0}^{n-1}(-1)^{k}\phi(ze^{-2k\pi i/n}),$$

$$F(X)=X^{p}$$

and remember the identity

$$\frac{1}{n}\sum_{k=0}^{n-1}(-1)^k\frac{1+ze^{-2k\pi i/n}}{1-ze^{-2k\pi i/n}}$$

$$=\frac{1}{n}\sum_{k=0}^{n-1}(-1)^k\left(1+2\sum_{h=1}^{\infty}z^h e^{-2hk\pi i/n}\right)$$

$$=2\sum_{\mu=0}^{\infty}z^{(\mu+1/2)n}=\frac{2z^{n/2}}{1-z^n},$$

in which the last member shows, in particular, that $\mathcal{L}[(1+z)/(1-z)]$ is single-valued in $z^{n/2}$ but not in higher power of z. The estimate then becomes after change of variables

$$\int_{-\pi}^{\pi}\left|\frac{d^\nu}{d(re^{i\theta})^\nu}\frac{2(re^{i\theta})^{n/2}}{1-(re^{i\theta})^n}\right|^p d\theta$$

$$=\int_{-\pi}^{\pi}\left|\frac{\partial^\nu}{\partial r^\nu}\frac{2r^{n/2}e^{i\theta}}{1-r^n e^{i\theta}}\right|^p d\theta. \qquad \square$$

REMARK. For $F(X) = X^p$, the inequality on convex functions used in the proof of Theorem 28.1 reduces to Hölder's. More particularly, it degenerates for the case $p = 1$ to a trivial one, as seen in the proof of Theorem 28.1.

On the other hand, it may be noted that, in Theorem 28. 5, the operation $z^\nu(d/dz)^\nu$ involved in $\mathcal{L}$ can be replaced, for instance, by $(z\,d/dz)^{\nu} = (d/d\log z)^{\nu}$, whence follows an analogue theorem.

As the next example illustrating a consequence of Theorems 28.1 and 28.4, we mention here the following Theorem:

THEOREM 28.6. *Let* $p \geq 1$. *Then, for any* $\phi \in P(0)$, *we have*

$$\int_{-\pi}^{\pi} \left| \int_0^r \phi(te^{i\theta})\,dt \right|^p d\theta$$

$$\leq \int_{-\pi}^{\pi} \left| 2\log\frac{1}{1 - re^{i\theta}} - re^{i\theta} \right| d\theta \quad (r \in (0, 1)).$$

The equality sign holds for any fixed r, *if and only if* ϕ *is of the form* $(1 + \varepsilon z)/(1 - \varepsilon z)$ *with* $|\varepsilon| = 1$.

Proof. We may put in Theorems 28.1 and 27.4

$$\mathcal{L}[\phi(z)] = \frac{1}{z}\int_0^z \phi(\zeta)\,d\zeta = \frac{1}{r}\int_0^r \phi(te^{i\theta})\,dt$$

$$(z = re^{i\theta}),$$

$$F(X) = X^p,$$

whence follows the desired estimation after reducing both members of that from Theorem 28.1 by $1/r^p$. In applying Theorem 28.4, we have only to verify that

$$\mathscr{L}\left[\frac{1+z}{1-z}\right] = -\frac{1}{z}(2\log(1-z)+z)$$

does not vanish in $\{0 < |z| < 1\}$. It is a simple consequence of a Theorem of Noshiro [1] and Wolff [2]; cf. also Theorem 15. 3 and cf. also Ozaki [1]. □

§ 29. Mean distortions of fractional integral

It is well-known that the fundamental operations, integration and differentiation, in ordinary calculus can be analytically interpolated to those of any real order. In the present section we shall show that the notion of the fractional calculus is also useful in dealing with distortion theorems in the Carathéodory class $P(0)$; cf. Komatu [4].

Though the results which will be derived below are essentially involved in general theorems in the preceding section, we may regard them as an interpolating generalization of some illustrating theorems given there. It will be of some interest to point out concrete cases where the estimates in the distortion inequalities are expressed in terms of integrals of elementary functions.

Since the unit disk E is a convex domain, the fractional integration and differentiation can be defined uniquely for any analytic function holomorphic in E with respect to any reference point in the domain. We suppose here that the reference point is always at the origin.

Let now q be any positive real number and $\mathcal{D}^{-q}$ denote the integration of order q. Then, the operator

$$\mathcal{L} \equiv \mathcal{L}_z : \quad \mathcal{L} = z^{-q} \mathcal{D}^{-q}$$

is linear and homogeneous of order zero, that is, for any constants a and b the function $\mathcal{L}[a\phi(z)]$ coincides after substitution $z \mid bz$ with $a\mathcal{L}[\phi(bz)]$. In fact, we have by definition

$$\mathcal{D}^{-q}\phi(z) = \frac{1}{\Gamma(q)} \int_0^z (z-\zeta)^{q-1} \phi(\zeta)\, d\zeta$$

where the branch of $(z-\zeta)^{q-1} \equiv \exp((q-1)\log(z-\zeta))$ is determined by taking the principal value of the logarithm and the integration is to be taken along the rectilinear segment from 0 to z. Putting $\zeta = tz$, we get

$$\mathcal{D}^{-q}\phi(z) = \frac{z^q}{\Gamma(q)} \int_0^1 (1-t)^{q-1} \phi(tz)\, dt .$$

Hence, it follows

$$[Z^{-q} \mathcal{D}_Z^{-q} a\phi(Z)]^{Z=bz} = \frac{a}{\Gamma(q)} \int_0^1 (1-t)^{q-1} \phi(tbz)\, dt$$

$$= az^{-q} \mathcal{D}_Z^{-q} \phi(bz).$$

Let next p be any positive real number and $\mathcal{D}^p$ denote the differentiation of order p. Then, the operator $\mathcal{L} = z^p \mathcal{D}^p$ is also linear and homogeneous of order zero. In fact, let $n = -[-p]$ be least integer not less than p and $s = n - p$; $0 \leqq s < 1$. Then, we have by definition

$$\mathcal{D}^p = \mathcal{D}^n \mathcal{D}^{-s}, \qquad \mathcal{D}^n = \frac{d^n}{dz^n}.$$

Consequently, it follows

$$[Z^p \mathcal{D}_Z^p a\phi(Z)]^{Z=bz} = [Z^p \mathcal{D}_Z^n \mathcal{D}_Z^{-s} a\phi(Z)]^{Z=bz}$$

$$= a(bz)^p b^{-n} \mathcal{D}_z(bz)^s z^{-s} \phi_z^{-s} \phi(bz)$$

$$= ab^{p-n+s} z^p \mathcal{D}_z^n \mathcal{D}_z^{-s} \phi(bz)$$

$$= az^p \mathcal{D}_z^p \phi(bz).$$

Since $\mathcal{D}^0 \equiv \mathcal{D}^{-0}$ is interpreted as the identity, the last relation remains valid obviously in case where p is an integer.

REMARK. When p is a positive integer, $z^p \mathcal{D}^p$ becomes a linear combination of $\{(d/d\log z)^\nu\}_{\nu=1}^p$; cf. § 13.

We begin with a mean distortion for fractional integral of $\phi \in P(0)$.

THEOREM 29.1. *Let* $q > 0$ *and* $\lambda \geqq 1$ *be any real numbers. Then, for any* $\phi \in P(0)$ *and* $r \in [0, 1)$, *we have*

$$\int_{-\pi}^{\pi} |\mathcal{D}^{-q}\phi(re^{i\theta})|^{\lambda} d\theta$$

$$\leq \frac{r^{q\lambda}}{\Gamma(q)} \int_{-\pi}^{\pi} \left| \int_I (1-t)^{q-1} \frac{1+tre^{i\theta}}{1-tre^{i\theta}} dt \right|^{\lambda} d\theta.$$

For any fixed $r \in (0, 1)$, *the equality sign holds if and only if* ϕ *is of the form* $\phi^*(\varepsilon z) = (1+\varepsilon z)/(1-\varepsilon z)$ *with* $|\varepsilon| = 1$.

Proof. By making use of Herglotz representation of ϕ, we get

$$\mathcal{D}^{-q}\phi(z) = \frac{z^q}{\Gamma(q)} \int_I (1-t)^{q-1} \phi(tz)\, dt$$

$$= \frac{z^q}{\Gamma(q)} \int_I (1-t)^{q-1} dt \int_{-\pi}^{\pi} \phi^*(e^{-i\varphi} tz)\, d\rho(\varphi)$$

and hence

$$\int_{-\pi}^{\pi} |\mathcal{D}^{-q}\phi(re^{i\theta})|^{\lambda}\, d\theta$$

$$= \frac{r^{q\lambda}}{\Gamma(q)^{\lambda}} \int_{-\pi}^{\pi} d\theta \left| \int_{-\pi}^{\pi} d\rho(\varphi) \right.$$

$$\left. \cdot \int_I (1-t)^{q-1} \phi^*(tre^{i(\theta-\varphi)})\, dt \right|^{\lambda}.$$

Since X^{λ} is an increasing convex function of $X \geqq 0$ and ρ is an increasing function with total variation equal to unity, the last relation yields

$$\int_{-\pi}^{\pi} |\mathcal{D}^{-q}\phi(re^{i\theta})|^{\lambda}\, d\theta$$

$$\leqq \frac{r^{q\lambda}}{\Gamma(q)^{\lambda}} \int_{-\pi}^{\pi} d\theta \int_{-\pi}^{\pi} d\rho(\varphi)$$

$$\cdot \left| \int_I (1-t)^{q-1} \phi^*(tre^{i(\theta-\varphi)})\, dt \right|^{\lambda}$$

$$= \frac{r^{q\lambda}}{\Gamma(q)^{\lambda}} \int_{-\pi}^{\pi} \left| \int_I (1-t)^{q-1} \phi^*(tre^{i\theta})\, dt \right|^{\lambda} d\theta.$$

This is the desired inequality. Concerning the equality sign, we first consider the case $\lambda = 1$. Then, any function ρ associated with an extremal function ϕ is characterized by the condition

$$\left| \int_{-\pi}^{\pi} d\rho(\varphi) \int_I (1-t)^{q-1} \phi^*(tre^{i(\theta-\varphi)})\, dt \right|$$

$$= \int_{-\pi}^{\pi} d\rho(\varphi) \left| \int_I (1-t)^{q-1} \phi^*(tre^{i(\theta-\varphi)})\, dt \right|$$

valid for every $\theta \in (-\pi, \pi]$, since its both members are continuous in θ. Define the function

$$\Psi(z) = \int_I (1-t)^{q-1} \phi^*(tz)\, dt$$

$$\left(\equiv \Gamma(q) z^{-q} \mathcal{D}^{-q} \phi^*(z)\right),$$

which is evidently holomorphic and of positive real part for $\{|z| \leqq r < 1\}$. The above condition is then equivalent to the requirement that for every value of θ the quantity $\Psi(re^{i(\theta-\varphi)})$ possesses the same argument for all $\varphi \in (-\pi, \pi]$ with $d\rho(\varphi) > 0$. But, it is shown that this quantity can never have the same argument for any distinct values of φ. In fact, suppose that $\Psi(ze^{-i\varphi})$ has the same argument at φ_0 and φ_1. Then, the function defined by

$$X(z) = \frac{\Psi(ze^{-i\varphi_1})}{\Psi(ze^{-i\varphi_0})}$$

is holomorphic throughout $\{|z| \leqq r\}$ and remains real along the circumference $\{|z| = r\}$. Hence, it must reduce to a constant, i. e., $X(z) \equiv X(0) = 1$. Since it holds

$$\left[\frac{d}{dz} \Psi(ze^{-i\varphi})\right]^{z=0} = e^{-i\varphi} \Psi'(0) = \frac{2}{q(q+1)} e^{-i\varphi},$$

φ_0 and φ_1 must coincide. Consequently, ρ remains constant except at a single jump with the height necessarily equal to unity. Thus, it is shown that the form of extremal function is given by $\phi^*(\varepsilon z)$ with $|\varepsilon| = 1$. Finally, in case of $\lambda > 1$, i. e., the case where X^λ is strictly convex, it is evident that the above condition is necessary. It is also directly obvious that the condition is sufficient. □

The statement in Theorem 29.1 remains true for $q = 0$, provided the estimate in the right-hand member is understood to be replaced by its limit as $q \to +0$. It reduces then trivially to

$$\lim_{q\to+0} \frac{r^{q\lambda}}{\Gamma(q)^{\lambda}} \int_{-\pi}^{\pi} \left| \int_I (1-t)^{q-1} \frac{1+tre^{i\theta}}{1-tre^{i\theta}} dt \right|^{\lambda} d\theta$$

$$= \int_{-\pi}^{\pi} \left| \frac{1+re^{i\theta}}{1-re^{i\theta}} \right|^{\lambda} d\theta.$$

We remark that the particular case $\lambda = 2$ of Theorem 29.1 can be verified alternatively in a simple way. In fact, let the Taylor expansion of ϕ be

$$\phi(z) = 1 + \sum_{\nu=1}^{\infty} c_{\nu} z^{\nu},$$

$$c_{\nu} = 2 \int_{-\pi}^{\pi} e^{-i\nu\varphi} d\rho(\varphi) \qquad (\nu = 1, 2, \ldots).$$

Since the operation $\mathcal{D}^{-q}$ is applicable termwise, we get

$$\mathcal{D}^{-q} \phi(z) = \frac{1}{\Gamma(1+q)} z^{q} + \sum_{\nu=1}^{\infty} c_{\nu} \frac{\nu!}{\Gamma(1+\nu+q)} z^{\nu+q},$$

$$\frac{1}{2\pi} \int_{-\pi}^{\pi} |\mathcal{D}^{-q} \phi(re^{i\theta})|^2 d\theta$$

$$= \frac{1}{\Gamma(1+q)^2} r^{2q} + \sum_{\nu=1}^{\infty} |c_\nu|^2 \frac{\nu!^2}{\Gamma(1+\nu q)^2} r^{2(\nu+q)}.$$

In view of $|c_\nu| \leq 2$ $(\nu \geq 1)$, there follows readily

$$\frac{1}{2\pi}\int_{-\pi}^{\pi} |\mathcal{D}^{-q}\phi(re^{i\theta})|^2 d\theta$$

$$\leq \frac{1}{\Gamma(1+q)^2} r^{2q} + 4\sum_{\nu=1}^{\infty} \frac{\nu!^2}{\Gamma(1+\nu+q)^2} r^{2(\nu+q)},$$

which is equivalent to the desired inequality. The assertion on extremal functions is immediate.

We next give a corresponding mean distortion for fractional derivative of $\phi \in P(0)$.

THEOREM 29.2. *Let* $p > 0$ *and* $\lambda \geq 1$ *be any real numbers and* $n = -[-p]$, $p = n - s$. *Then, for any* $\phi \in P(0)$ *and* $r \in [0, 1)$, *we have*

$$\int_{-\pi}^{\pi} |\mathcal{D}^{p}\phi(re^{i\theta})|^\lambda d\theta$$

$$\leq \frac{s}{r^p}\int_{-\pi}^{\pi} \left| \int_I (1-t)^{s-1}\left(\frac{1}{\Gamma(1-p)} \frac{1+tre^{i\theta}}{1-tre^{i\theta}}\right.\right.$$

$$\left.\left. + 2\sum_{j=1}^{\infty} \frac{n!}{(n-j)!} \frac{1}{\Gamma(1+j-p)} \frac{(tre^{i\theta})^j}{(1-tre^{i\theta})^{j+1}}\right) dt \right|^\lambda d\theta.$$

The function $\phi^*(\varepsilon z)$ *with* $|\varepsilon| = 1$ *is always an extremal function for this estimation.*

Proof. Herglotz representation yields

$$\mathcal{D}^p \phi(z) = \int_{-\pi}^{\pi} \mathcal{D}^p \phi^*(e^{-i\varphi} z)\, d\rho(\varphi),$$

and hence

$$\int_{-\pi}^{\pi} |\mathcal{D}^p \phi(re^{i\theta})|^\lambda d\theta$$

$$= \int_{-\pi}^{\pi} d\theta \left| \int_{-\pi}^{\pi} \mathcal{D}^p \phi^*(e^{-i\varphi} re^{i\theta})\, d\rho(\varphi) \right|^\lambda$$

$$\leq \int_{-\pi}^{\pi} d\theta \int_{-\pi}^{\pi} |\mathcal{D}^p \phi^*(re^{i(\theta-\varphi)})|^\lambda d\rho\varphi$$

$$= \int_{-\pi}^{\pi} |\mathcal{D}^p \phi^*(re^{i\theta})|^\lambda d\theta.$$

It remains only to compute the last estimate in explicit form. Based on the definition of fractional derivative, we have

$$\mathcal{D}^p \phi^*(z) = \mathcal{D}^n \frac{z^s}{\Gamma(s)} \int_I (1-t)^{s-1} \phi^*(tz)\, dt$$

$$= s \sum_{j=0}^{n} \binom{n}{j} \frac{z^{s-n+j}}{\Gamma(s-n+j+1)}$$

$$\cdot \int_I (1-t)^{s-1} t^j \phi^*(tz)\, dt$$

$$= s\int_I (1-t)^{s-1}\left(\frac{z^{-p}}{\Gamma(1-p)}\phi^*(tz) + \sum_{j=1}^{n}\frac{n!\,t^j z^{-p+j}}{(n-j)!\,j!\,\Gamma(1+j-p)}\phi^{*(j)}(tz)\right)dt$$

Thus, by remembering

$$\phi^*(tz) = \frac{1+tz}{1-tz}, \qquad \phi^{*(j)}(tz) = \frac{j!\,2}{(1-tz)^{j+1}} \qquad (j \geqq 1),$$

the desired form follows immediately by direct substitution.

□

If, in particular, p is equal to a natural number n, the estimate in the right-hand member in Theoem 28.2 becomes briefly

$$\lim_{p\to n-0}\int_{-\pi}^{\pi} |\mathcal{D}^p\phi^*(re^{i\theta})|^\lambda\, d\theta$$

$$= \int_{-\pi}^{\pi} |\mathcal{D}^n\phi^*(re^{i\theta})|^\lambda\, d\theta$$

$$= (n!\,2)^\lambda\, 2\int_0^{\pi} \frac{d\theta}{|1-re^{i\theta}|^{(n+1)\lambda}}.$$

This is the case previously dealt with from the more general viewpoint; cf. Theorem 28.3.

Finally, we supplement a remark on the characterization

of extremal functions in Theorem 29.2, though it will not be quite complete. Any function $\phi^*(\varepsilon z)$ with $|\varepsilon| = 1$ is surely an extremal function as mentioned in the Theorem. Now it depends only on the behavior of the elementary function $\phi^*(z)$ whether there exists besides an extremal function or not. In fact, in case $\lambda = 1$, any function ρ associated with an extremal function ϕ is characterized by the relation

$$\left| \int_{-\pi}^{\pi} \mathcal{D}^p \phi^*(re^{i(\theta-\varphi)})\, d\rho(\varphi) \right|$$

$$= \int_{-\pi}^{\pi} |\mathcal{D}^p \phi^*(re^{i(\theta-\varphi)})|\, d\rho(\varphi)$$

valid for every $\theta \in (-\pi, \pi]$, since its both members are continuous in θ. This condition is equivalent to the requirement that for every θ the quantity $\mathcal{D}^p \phi^*(re^{i(\theta-\varphi)})$ possesses the same argument for all φ with $d\rho(\varphi) > 0$ over $(-\pi, \pi]$.

Now, if it could be shown that $\mathcal{D}^p \phi^*(z)$ does not vanish for $0 < |z| \leq r$, we can prove that $\mathcal{D}\phi^*(re^{i(\theta-\varphi)})$ has never the same argument for any distinct values of φ. In fact, the function $z^p \mathcal{D}^p \phi^*(z)$ is holomorphic for $\{|z| \leq r\}$. Suppose that $z^p \mathcal{D}^p \phi^*(ze^{-i\varphi})$ has the same argument at φ_0 and φ_1. Then, the quotient

$$Y(z) = \frac{\mathcal{D}^p \phi^*(ze^{-i\varphi_0})}{\mathcal{D}^p \phi^*(ze^{-i\varphi_1})}$$

is holomorphic throughout $\{|z| \leqq r\}$ and remains real along $\{|z| = r\}$. Hence, it must reduce to a constant, i. e., $Y(z) \equiv Y(0) = 1$. Consequently, in view of

$$\left[\frac{d}{dz} z^{p} \mathcal{D}^{p} \phi^{*}(ze^{i\varphi})\right]^{z=0} = \frac{2e^{-i\varphi}}{\Gamma(2-p)},$$

we conclude that φ_0 and φ_1 cannot differ provided $p \neq 2, 3, \ldots$. But, if p is not a positive integer, then we get

$$\mathcal{D}^{p}\phi^{*}(ze^{-i\varphi}) = \frac{p!\,2e^{-ip\varphi}}{(1-ze^{-i\varphi})^{p+1}}$$

and hence φ_0 and φ_1 cannot differ surely.

Consequently, it is verified that extremal functions in Theorem 29.2 for $\lambda = 1$ and a fortiori for $\lambda > 1$ must be of the form $\phi^{*}(\varepsilon z)$ with $|\varepsilon| = 1$, provided that $\mathcal{D}^{p}\phi^{*}(z)$ does not vanish for $\{0 < |z| \leqq r\}$. Further, it has been shown that the last-mentioned condition on $\mathcal{D}^{p}\phi^{*}(z)$ is really satisfied for any positive integer p.

It may be here noticed that, for any $p > 0$, the condition $\mathcal{D}^{p}\phi^{*}(z) \neq 0$ for $0 < |z| \leqq r$ is naturally fulfilled, provided r is small enough.

On the other hand, the particular case of $\lambda = 2$ can be dealt with directly in a similar manner as before. In fact, putting

$$\phi(z) = 1 + \sum_{\nu=1}^{\infty} c_\nu z^\nu, \qquad c_\nu = \int_{-\pi}^{\pi} e^{-i\nu\varphi} d\rho(\varphi) \quad (\nu \geqq 1),$$

we get

$$\mathcal{D}^p \phi(z) = \frac{1}{\Gamma(1-p)} z^{-p} + \sum_{\nu=1}^{\infty} c_\nu \frac{\nu!}{\Gamma(1+\nu-p)} z^{\nu-p},$$

$$\frac{1}{2\pi}\int_{-\pi}^{\pi} |\mathcal{D}^p \phi^*(re^{i\theta})|^2 d\theta$$

$$= \frac{1}{\Gamma(1-p)^2} r^{-2p} + \sum_{\nu=1}^{\infty} |c_\nu|^2 \frac{\nu!^2}{\Gamma(1+\nu-p)^2} r^{2(\nu-p)}.$$

Consequently, we obtain the mean distortion in the form

$$\frac{1}{2\pi}\int_{-\pi}^{\pi} |\mathcal{D}^p \phi(re^{i\theta})|^2 d\theta$$

$$\leqq \frac{1}{\Gamma(1-p)^2} r^{-2p} + 4 \sum_{\nu=1}^{\infty} \frac{\nu!^2}{\Gamma(1+\nu-p)^2} r^{2(\nu-p)}.$$

The extremal functions in this case are only those of the form $\phi^*(\varepsilon z)$ with $|\varepsilon| = 1$, because even a single relation $|c_\nu| = 2$ for some ν implies $\phi(z) = \phi^*(\varepsilon z)$ with $\varepsilon^\nu = c_\nu/2$.

§ 30. Angular derivative

With respect to the notion of angular derivative, its germination is found in the theorem of Julia [1] generalizing the theorem of Schwarz [1] as well as in the theorem of Wolff [1], which was introduced at first for functions f holomorphic and satisfying Re $f(z) > 0$ in the half-plane {Re $z > 0$}. However, its importance has been generally accepted by a theorem of Nevanlinna [2, 3, 4], F. & R. [1] as well as in the theorems of Carathéodory [1] and of Landau-Valiron [1] both published almost simultaneously.

We take, in the following lines, the unit disk E as the basic domain and prepare for the present section the fundamental theorem on angular derivative as a lemma. We reproduce here an outline of the theory on angular derivative for the sake of self-containedness of this book; cf. Komatu [2, 3] and cf. also Pommerenke [1], p. 302 et seq.

We now begin with a theorem of Carathéodory [1] in the form supplemented by Herzig [1]; cf. also Wolff [1], Dinghas [1].

LEMMA 30.1. *Let f be an analytic function holomorphic in E and satisfying there $|f(z)| < 1$. Then there exists a positive constant D, eventually equal to $+\infty$, called the angular derivative of f at 1, such that*

$$\frac{1 - |f(z)|^2}{|1 - f(z)|^2} \geqq \frac{1}{D} \frac{1 - |z|^2}{|1 - z|^2};$$

hereafter D will be understood to denote the least possible constant satisfying this inequality. Then, the equality holds at a point if and only if f is a linear function defined by

$$\frac{1+f(z)}{1-f(z)} = \frac{1}{D}\frac{1+z}{1-z} + i\beta, \quad \beta = \mathrm{Im}\frac{1+f(0)}{1-f(0)};$$

in this case D is necessarily finite. In general, the function satisfies the limit relation

$$(30.1) \qquad \lim_{z\to 1} \frac{1-f(z)}{1-z} = D,$$

and if D is finite, the successive derivatives of f *satisfy the limit relations*

$$(30.2) \qquad \lim_{z\to 1} f'(z) = D$$

and

$$(30.3) \qquad (1-z)^{n-1} f^{(n)}(z) = 0 \quad (n = 2, 3, \dots);$$

here the integral power $n-1$ of $1-z$ is best possible, and both (30.2) *and* (30.3) *hold uniformly as z tends to* 1 *along any path through a Stolz angle in E with the vertex at* 1.

Proof. Since $(1+f)/(1-f)$ is holomorphic and possessing positive real part in E, the Herglotz representation yields

$$\frac{1+f(z)}{1-f(z)} = \int_{-\pi}^{\pi} \frac{e^{it}+z}{e^{it}-z}\, d\tau(t) + i\,\mathrm{Im}\frac{1+f(0)}{1-f(0)},$$

τ being a real-valued increasing function defined for $[-\pi, \pi]$ with total variation equal to Re $((1 + f(0))/(1 - f(0)))$.

It will be convenient to assume that τ is continued beyond the original interval of definition by the condition such that $\tau(t) - \text{Re}\,((1 + f(0))/(1 - f(0)))\,t/2\pi$ has 2π as its period. By separating the real part, and noting that the equality sign holds if f is the linear function, we get

$$\frac{1 - |f(z)|^2}{|1 - f(z)|^2} = \int_{-\pi}^{\pi} \frac{1 - |z|^2}{|e^{it} - z|^2}\, d\tau(t)$$

$$\geq \int_{-0}^{+0} \frac{1 - |z|^2}{|e^{it} - z|^2}\, d\tau(t)$$

$$= \frac{1}{D} \frac{1 - |z|^2}{|1 - z|^2},$$

where D is given by

$$\frac{1}{D} = \tau(+0) - \tau(-0), \tag{30.4}$$

in view of the continuation of τ mentioned above. In particular, we have

$$0 \leq \frac{1}{D} \leq \text{Re}\,\frac{1 + f(0)}{1 - f(0)}$$

$$= \int_{-\pi}^{\pi} d\tau(t) < +\infty,$$

that is,

$$0 < D \leqq +\infty.$$

The equality sign in the above inequality appears if and only if τ remains constant in $[-\pi, -0) \cup (+0, \pi]$. Hence, the extremal function must be a linear function given in the Theorem. Next, for any f under consideration, it holds

$$\frac{1}{1 - f(z)} = \int_{-\pi}^{\pi} \frac{e^{it}}{e^{it} - z} d\tau(t) - \frac{\overline{f(0)}}{1 - \overline{f(0)}}$$

$$= \frac{1}{D} \frac{z}{1 - z} + \left(\int_{-\pi}^{-0} + \int_{+0}^{\delta} \right) \frac{e^{it}}{e^{it} - z} d\tau(t) - \frac{\overline{f(0)}}{1 - \overline{f(0)}}.$$

For z tending to 1 along a Stolz path, an estimation of the form $|1 - z| / (1 - |z|) \leqq K < +\infty$ holds. Therefore, for any $\delta \in (0, \pi)$, we obtain

$$\left| \frac{1 - z}{1 - f(z)} - \frac{1}{D} \right|$$

$$\leqq K \left(\int_{-\delta}^{-0} + \int_{+0}^{\delta} \right) d\tau(t)$$

$$+ |1 - z| \left(\left(\int_{-\pi}^{-\delta} + \int_{\delta}^{\pi} \right) \frac{1}{|e^{it} - z|} d\tau(t) + \left| \frac{f(0)}{1 - f(0)} \right| \right).$$

Consequently, taking δ small enough and then letting z tend to 1 along a Stolz path, it follows that the limit relation

$$\frac{1 - z}{1 - f(z)} \to \frac{1}{D}, \quad \text{i. e.,} \quad \frac{1 - f(z)}{1 - z} \to D$$

holds uniformly. In quite a similar way, it follows from the derived formula

$$\frac{f'(z)}{(1-f(z))^2} = \int_{-\pi}^{\pi} \frac{e^{it}}{(e^{it}-z)^2}\, d\tau(t)$$

that the limit relation

$$\left(\frac{f(z)}{1-f(z)}\right)^2 f'(z) \to \frac{1}{D}$$

holds uniformly. Hence, by combining with the relation established above, we obtain the relation

$$f'(z) \to D$$

provided $D \neq \infty$, valid uniformly. Finally, suppose $n \leqq 2$ and $D < +\infty$. The Herglotz representation for $(1+f)/(1-f)$ or equivalently for $1/(1-f)$ becomes after differentiating n times

$$\frac{d^n}{dz^n}\frac{1}{1-f(z)}$$

$$= n! \int_{-\pi}^{\pi} \frac{e^{it}}{(e^{it}-z)^{n+1}}\, d\tau(t)$$

$$= \frac{n!}{D}\frac{1}{(1-z)^{n+1}} \quad n! \left(\int_{-\pi}^{-0} + \int_{+0}^{\pi}\right) \frac{e^{it}}{(e^{it}-z)^{n+1}}\, d\tau(t).$$

Hence, we conclude by means of a similar argument, as above, that the limit relation

$$(1-z)^{n+1}\frac{d^n}{dz^n}\frac{1}{1-f(z)}\to\frac{n!}{D}$$

holds uniformly for Stolz approach. Now, it is readily verified by induction that the n-th derivative of $1/(1-f(z)$ with rspect to z is a polynomial of n arguments

$$\omega_\nu\equiv\frac{f^{(\nu)}(z)}{1-f(z)}\qquad(\nu=1,\ldots,n)$$

multiplied by $1/(1-f(z))$, coefficients of the polynomial being nuerical constants independent of f. Further, if ω_ν is regarded as a quantity of degree ν, respectively, then this polynomial is homogeneous with degree n. In other words, the n-th derivative in question is of the form

$$\frac{d^n}{dz^n}\frac{1}{1-f(z)}$$

(30.5)

$$=\frac{1}{1-f(z)}\sum_{\substack{\nu_1,\ldots,\nu_k\leqq 1\\ \nu_1+\ldots+\nu_k=n}}\gamma^{(n)}_{\nu_1\ldots\nu_k}\prod_{\kappa=1}^{k}\frac{f^{(\nu_k)}(z)}{1-f(z)},$$

the γ's being numerical constants independent of f. Evidently, we obtain, in particular,

$$\gamma^{(n)}_n=1\quad\text{and}\quad\gamma^{(n)}_{1\ldots1}=n!.$$

In order to prove the last part of the theorem by induction, we first deal with the case $n = 2$ directly. For this case we have

$$\frac{d^2}{dz^2}\frac{1}{1-f(z)} = \frac{f''(z)}{(1-f(z))^2} + \frac{2f'(z)^2}{(1-f(z))^3}.$$

Consequently, in view of the limit relation established just above for general $n \geq 2$, it follows that

$$\frac{2}{D} = \lim_{z\to 1}(1-z)^3\frac{d^2}{dz^2}\frac{1}{1-f(z)}$$

$$= \lim_{z\to 1}\left(\left(\frac{1-z}{1-f(z)}\right)^2(1-z)f''(z) + 2\left(\frac{1-z}{1-f(z)}\right)^3 f'(z)\right)$$

$$= \frac{1}{D^2}\lim_{z\to 1}(1-z)f''(z) + 2\frac{1}{D^3}D^2,$$

i. e.,

$$\lim_{z\to 1}(1-z)f''(z) = 0$$

holds uniformly for Stolz approach. Suppose now

$$\lim_{z\to 1}(1-z)^{\nu-1}f^{(\nu)}(z) = 0 \quad \text{for } \nu = 2, \ldots, n-1$$

uniformly for Stolz approach. Then, we obtain

$$\frac{n!}{D} = \lim_{z\to 1}(1-z)^{n+1}\frac{d^n}{dz^n}\frac{1}{1-f(z)}$$

$$= \lim_{z\to 1} \Bigg(\frac{1-z}{1-f(z)} \sum_{\substack{\nu_1,\dots,\nu_k \geqq 1 \\ \nu_1+\dots+\nu_k = n}} \tau_{\nu_1\dots\nu_k} \prod_{\kappa=1}^{k} \frac{1-z}{1-f(z)} (1-z)^{\nu_\kappa - 1} f^{(\nu_\kappa)}(z) \Bigg)$$

$$= \frac{1}{D} \lim_{z\to 1} \Big(\frac{1}{D} (1-z)^{n-1} f^{(n)}(z) + n! \frac{1}{D^n} f'(z)^n \Big)$$

$$= \frac{1}{D^2} \lim_{z\to 1} (1-z)^{n-1} f^{(n)}(z) + n! \frac{1}{D^{n+1}} D^n .$$

This shows that the desired relation

$$\lim_{z\to 1} (1-z)^{n-1} f^{(n)}(z) = 0$$

holds uniformly in general for $n \geq 2$ as z tends to 1 along any Stolz path. In conclusion, it is noticed that, for instance, the branch of

$$g(z) = 1 - \frac{4}{\zeta + \zeta^m + 2} \quad \text{with} \quad \zeta \equiv \frac{1+z}{1-z}, \quad 0 < m < 1,$$

determined by $g(0) = 0$ satisfies the whole condition imposed in the present Lemma. Its angular derivative at $z = 1$ has a finite value $D = 2$. On the other hand, it is verified that its n-th derivative $g^{(n)}(z)$ grows near $z = 1$ with the order of $(1-z)^{-n+2-m}$. Since in this wxample m can be taken arbitrarily near the unity, the exponent n - 1 of $1 - z$ in the expression for $(1 - z)^{n-1} f^{(n)}(z)$ cannot be ameriorated.

On the way of the proof of Lemma 30.1, we have seen the relation (30.4). This shows, in particular, a necessary and sufficient condition for the angular derivative D to be finite, is that the function τ is discontinuous at 0, where τ is defined by

$$\tau(t) = \lim_{R\to 1-0} \frac{1}{2\pi} \int_{-\pi}^{t} \frac{1 - |f(Re^{i\varphi})|^2}{|1 - f(Re^{i\varphi})|^2} d\varphi.$$

REMARK. It would be remarked that (30.5) gives

$$\left[\frac{d^n}{dz^n} \frac{1}{1 - f(z)} \right]^{z=0}$$

$$= \frac{1}{1 - f(0)} \sum_{\substack{\nu_1, \ldots, \nu_k \geq 1 \\ \nu_1 + \ldots + \nu_k = n}} \gamma^{(n)}_{\nu_1 \ldots \nu_k} \prod_{\kappa=1}^{k} \frac{f^{(\nu_\kappa)}(0)}{1 - f(0)}$$

On the other hand, if f vanishes at the origin, then we may replace here $1/(1 - f(z))$ in the left-hand member by the quantity $\sum_{\nu=0}^{n} f(z)^\nu$. Thus, in this case, we obtain an alternative relation generating the γ's:

$$\sum_{\nu=1}^{n}\left[\frac{d^n}{dz^n} f(z)^{\nu}\right]^{z=0} = \sum_{\substack{\nu_1,\dots,\nu_k \geq 1 \\ \nu_1+\dots+\nu_k = n}} \gamma^{(n)}_{\nu_1\dots\nu_k} \prod_{\kappa=1}^{k} f^{(\nu_\kappa)}(0).$$

Since the last identity remains valid for any analytic function f satisfying $|f| < 1$ and vanishing at the origin, it will serve the actual determination of the numerical values of the γ's.

In the Lemma 30.1, if $f(z)$ is subject to the condition $f(0) = 0$, it follows that its angular derivative D at 1 satisfies $D \geq 1$ and that the equality holds if and only if $f(z) \equiv z$. Under the additional condition that $f'(0)$ has a preassigned value, a more precise estimation was derived by Unkelbach [1] and soon later reproved by Herzig [1]. Since in view of Schwarz's lemma we have always $|f'(0)| \leqq 1$ with the equality valid for $f(z) \equiv f'(0)z$ alone, the case $|f'(0)| < 1$ only is essential. The result of Unkelbach and Herzig states that

$$D \geqq 2\frac{1 - \operatorname{Re} f'(0)}{1 - |f'(0)|^2} \quad (> 1)$$

and that the equality sign appears if and only if

$$f(z) = z\frac{(1 - f'(0))z + f'(0)(1 - \overline{f'(0)})}{\overline{f'(0)}(1 - f'(0))z + 1 - \overline{f'(0)}}.$$

Now, we shall show that this result can be generalized as in the following manner; cf. Komatu[3]:

THEOREM 30.1. *Let f be an analytic function holomorphic and satisfying $|f(z)| < 1$ in E. Then, we have an inequality*

$$\left| \frac{f'(z)}{(1-f(z))^2} - \frac{1}{D}\frac{1}{(1-z)^2} \right|$$

$$\leqq \left(\mathrm{Re}\,\frac{1+f(0)}{1-f(0)} - \frac{1}{D}\right)\frac{1}{(1-|z|)^2}$$

valid for every $z \in E$. At any assigned $z_0 = |z_0|e^{i\theta_0} \in E$, the equality sign appears if and only if f is a rational function satisfying

$$\frac{1}{1-f(z)}$$

$$= \frac{1}{D}\frac{1}{1-z} + \left(\mathrm{Re}\,\frac{1+f(0)}{1-f(0)} - \frac{1}{D}\right)\frac{e^{i\theta_0}}{e^{i\theta_0}-z} - \frac{\overline{f(0)}}{1-\overline{f(0)}}.$$

When the value of $f'(z_0)/(1-f(z_0))^2$ is preassigned, the value of D for the extremal function may be determined by the equation

$$\frac{f'(z_0)}{(1-f(z_0))^2} = \frac{1}{D}\frac{1}{(1-z_0)^2} + \left(\mathrm{Re}\,\frac{1+f(0)}{1-f(0)} - \frac{1}{D}\right)\frac{e^{i\theta_0}}{(e^{i\theta_0}-z_0)^2}.$$

Proof. From a relation used in the proof of Lemma 30.1, we get

$$\frac{f'(z)}{(1-f(z))^2} = \frac{1}{D}\frac{1}{(1-z)^2} + \int_C \frac{e^{it}}{(e^{it}-z)^2}\, d\tau(t),$$

C denoting $\{(-\pi, \pi] - [-0, +0\}$. Thus, the evident relations

$$\left|\frac{e^{it}}{(e^{it}-z)^2}\right| \leqq \frac{1}{(1-|z|)^2},$$

$$d\tau(t) \geqq 0, \qquad \int_C d\tau(t) = \mathrm{Re}\,\frac{1+f(0)}{1-f(0)} - \frac{1}{D}$$

imply readily the desired result.

The estimation of D by Unkelbach and Herzig is regarded as a special case of the Theorem 30.1. We have only to put $f(0) = 0$ and $z = 0$. In fact, we then obtain

$$\left| f'(0) - \frac{1}{D} \right| \leqq 1 - \frac{1}{D}, \quad \text{i. e.,} \quad D \geqq 2\,\frac{1-\mathrm{Re}\, f'(0)}{1-|f'(0)|^2}.$$

For $f'(0)$ preassigned, the extremal function f is defined by

$$\frac{1}{1-f(z)} = \frac{1}{D}\frac{1}{1-z} + \left(1-\frac{1}{D}\right)\frac{e^{i\theta_0}}{e^{i\theta_0}-z}$$

with $f'(0) = 1/D + (1 - 1/D)e^{-i\theta_0}$. Hence, inserting the values

$$\frac{1}{D} = \frac{1 - f'(0)\overline{f'(0)}}{2 - f'(0) - \overline{f'(0)}}$$

and

$$e^{i\theta_0} = \frac{f'(0) - 1/D}{1 - 1/D} = -\frac{1 - \overline{f'(0)}}{1 - f'(0)},$$

we really obtain

$$f(z) = z\,\frac{(1 - f(0))z + f'(0)(1 - \overline{f'(0)})}{f'(0)(1 - f'(0))z + 1 - f'(0)}.$$

Likewise we can derive an inequality involvingthe higher order derivatives:

THEOREM 30.2. *For* f *considered in Theorem* 30.1, *we have, for any integer* $n \geqq 1$, *an inequality*

$$\left|\frac{1}{n!}\frac{d^n}{dz^n}\frac{1}{1 - f(z)} - \frac{1}{D}\frac{1}{(1 - z)^{n+1}}\right|$$

$$\leqq \left(\mathrm{Re}\,\frac{1 + f(0)}{1 - f(0)} - \frac{1}{D}\right)\frac{1}{(1 - |z|)^{n+1}}$$

At any assigned $z_0 = |z_0| e^{i\theta_0} \in E$, *the equality sign appears if and only if* f *satisfies*

$$\frac{1}{1-f(z)} = \frac{1}{D}\frac{1}{1-z} + \left(\mathrm{Re}\frac{1+f(0)}{1-f(0)} - \frac{1}{D}\right)\frac{e^{i\theta_0}}{e^{i\theta_0}-z} - \frac{\overline{f(0)}}{1-f(0)}$$

when $z_0 \neq 0$, *while it satisfies*

$$\frac{1}{1-f(z)} = \frac{1}{D}\frac{1}{1-z} + \sum_{j=1}^{n-1}\tau_j\frac{e^{i(t_0+2j\pi/n)}}{e^{i(t_0+2j\pi/n)}-z} - \frac{\overline{f(0)}}{1-f(0)}$$

when $z_0 = 0$, *where* t_0 *is any real number and the* τ'*s denote arbitrary real numbers subject to the conditions*

$$\tau_j \geq 0 \quad (j = 0, \ldots, n-1) \quad \text{and} \quad \sum_{j=0}^{n-1}\tau_j = \mathrm{Re}\frac{1+f(0)}{1-f(0)} - \frac{1}{D}$$

Proof. We can proceed similarly as for Theorem 30.2. In fact, we have only to remember

$$\frac{1}{n!}\frac{d^n}{dz^n}\frac{1}{1-f(z)} = \frac{1}{D}\frac{1}{(1-z)^{n+1}} + \left(\int_{-\pi}^{-0} + \int_{+0}^{\pi}\right)\frac{e^{it}}{(e^{it}-z)^{n+1}}\,d\tau(t). \qquad \square$$

Now, we shall observe the fractional differentiation. The fractional derivative of order p, i. e., the fractional integral of order $-p$ is defined by

$$\mathcal{D}^{p} = \mathcal{J}^{-p} \qquad \left(\mathcal{D} \equiv \frac{d}{dz} \right).$$

Then, the relation (30.3) together with (30.1) is generalized as follows; cf. Komatu [2]:

THEOREM 30.3. *Let f be an analytic function holomorphic and satisfying $|f(z)| < 1$ in E. Then, there exists a positive real constant D, called the fractional angular derivative of f and eventually equal to $+\infty$, such that the derivative of $f - 1$ of any real order p satisfies the limit relation*

$$\lim_{z \to 1} (z - 1)^{p-1} \mathcal{D}^{p}(f(z) - 1)$$

$$\equiv \lim_{z \to 1} \left((z - 1)^{p-1} \mathcal{D}^{p} f(z) - \frac{1}{\Gamma(1 - p)} \frac{1}{z - 1} \right)$$

$$= \frac{D}{\Gamma(2 - p)}$$

valid uniformly as z tends to 1 through any aggular region $\{|\arg(1 - z)| \leq \alpha < \pi/2\}$ in E. Here, it is supposed that D is finite when p is positive.

Proof. Suppose first $p = -q$ and $D \neq +\infty$. Since as $z \to 1$ along any Stolz path, we have in view of the lemma 30.1 the relation

$$f(1 + t(z - 1)) = (Dt + o(1))(z - 1)$$

valid uniformly in the wider sense for $t \in (0, 1)$. Hence, we get

$$\mathscr{D}^{-q}(f(z) - 1)$$

$$= \frac{(z-1)^q}{\Gamma(q)} \int_I (1 - t)^{q-1} (\mathscr{D}t + o(1))(z - 1)\, dt)$$

$$= \frac{(z-1)^{q+1}}{\Gamma(2+q)} (D + o(1)),$$

i. e.,

$$(z - 1)^{p-1} \mathscr{D}^p(f(z) - 1) = \frac{D}{\Gamma(2 - p)} + o(1).$$

Let $D = +\infty$. We write

$$\mathrm{Re}\,((z - 1)^{-q-1} \mathscr{D}^{-q}(f(z) - 1))$$

$$= \frac{1}{\Gamma(q)} \int_I (1 - t)^{q-1} t\, \mathrm{Re} \frac{f(1 + t(z - 1)) - 1}{t(z - 1)}\, dt.$$

Since, as $z \to 1$ along any Stolz path,

$$\mathrm{Re} \frac{f(1 + t(z - 1)) - 1}{t(z - 1)} \to +\infty$$

uniformly in the wider sense for $t \in (0, 1)$, the right hand member of the above equation also tends to $+\infty$. Consequently, the required limit relation holds also in this case. The case $p = 0$ is nothing but the first relation in Lemma 30.1. Next, we suppose $p > 0$ and $D \neq +\infty$. Since $1/\Gamma(2-p)$ with an integral value p is equal to 1 or 0 for $p = 1$ or $p \geqq 2$, respectively, the result has already be noticed in (30.2). Hence we may suppose that p is not an integer. Put $p = n - s$ with $n = [p] + 1$. Then, performing out the repeated differentiation, the defining equation of the pth derivative becomes

$$\mathscr{D}^p(f(z)-1)$$

$$= \frac{s(z-1)^{s-n}}{\Gamma(s-n+1)} \int_I (1-t)^{s-1}(f(1+t(z-1))-1)\,dt$$

$$+ s\sum_{\nu=0}^{n} \binom{n}{\nu} \frac{(z-1)^{s-n+\nu}}{\Gamma(s-n+\nu+1)}$$

$$\cdot \int_I (1-t)^{s-1} t^{\nu} f^{(\nu)}(1+t(z-1)\,dt\,.$$

In view of the known asymptotic behavior of $f^{(\nu)}$ with integral value of ν, we have

$$f^{(\nu)}(1+t(z-1))$$

$$= \begin{cases} 1+((Dt+o(1))(z-1) & (\nu = 0), \\ D+o(1) & (\nu = 1), \\ o(1)(z-1)^{1-\nu} & (\nu = 2, \ldots, n) \end{cases}$$

valid uniformly in the wider sense for $t \in (0, 1)$, provided $D \neq +\infty$. It follows that

$$\mathcal{D}^{p}(f(z) - 1)$$

$$= \frac{s(z-1)^{s-n}}{\Gamma(s-n+1)} \int_I (1-t)^{s-1}(Dt + o(1))(z-1)\,dt$$

$$+ \frac{sn(z-1)^{s-n+1}}{\Gamma(s-n+2)} \int_I (1-t)^{s-1} t (D + o(1))\,dt$$

$$+ \sum_{\nu=2}^{n} (z-1)^{s-n+\nu} o(1)(z-1)^{1-\nu}$$

$$= \frac{(z-1)^{s-n+1}}{\Gamma(s-n+2)} (D + o(1)),$$

i. e.,

$$(z-1)^{p-1} \mathcal{D}^{p}(f(z) - 1) = \frac{D}{\Gamma(2-p)} + o(1),$$

since we have $\mathcal{D}^{p} 1 = (z-1)^{-p}/(1-p)$.

As seen from the proof of Theorem 30.4, it is evident why the principal part in the asymptotic formula for $\mathcal{D}^{p} f$ or $\mathcal{D}^{p}(f-1)$ with non-integral value of p and integral value of $p \leq 1$ is exactly of order z^{-p+1} or $(z-1)^{-p+1}$, respectively, while that with integral value of $p \geqq 2$ is of lower order. In fact, $1/\Gamma(2 - p)$ is an entire function of p whose zero points, all simple, coincide just with integers $\geqq 2$.

References

Alexander, J. W.: [1] Functions which map the interior of the unit circle upon simple regions. Ann. Math. **17** (1915), 12-22.

Altintas, O.: [1] On the coefficients of functions majorized by univalent functions. Hacettepe Bull. Nat. Sci.-Eng. **10** (1981), 23-31.

Bernardi, S. D.: [1] Convex and starlike univalent functions. Trans. Amer. Math. Soc. **135** (1969) 429-446.

[2] Bibliography of schlicht functions. Marliner Publ. Co., Inc., Tampa, Florida (1982), x+353 pp.

Bieberbach, L.: [1] Über die Koeffizienten derjenigen Potenzreihen, welche eine schlichte Abbildung des Einheitskreises vermitteln. Stzber. Preuss. Akad. Wiss. Berlin (1916), 940-955.

Biernacki, M.: [1] Sur l'intégrale des fonctions univalentes. Bull. Acad. Polon. Sci., Sér. Sci. Math., Astr. et Phys. **8** (1960), 29-34.

Carathéodory, C..: [1] Über den Variabilitätsbereich der Koeffizienten von Potenzreihen, die gegebene Werte nicht annehmen. Math. Ann. **64** (1907), 95 -115.

[2] Über den Variabilitätsbereich der Fourierschen Konstanten von positiven harmonischen Funktionen. Rend. Palermo **32** (1911), 193-217.

[3] Über die Winkelderivierten von beschränkten analytischen Funktionen. Stzber. Preuss. Akad. (1929), 1-18.

Chichra, P. N.: [1] New subclasses of the class of close-to-convex functions. Proc. Amer. Math. Soc. **62** (1977), 37-43.

Clunie, J.: [1] On meromorphic schlicht functions. J. London Math. Soc. **34** (1959), 215-216.

de Branges, L.: [1] A proof of the Bieberbach conjecture. Acta Math. **154** (1985), 137-152.

Dinghas, A.: [1] Über das Phragmén-Lindelöfsche Prinzip und den Julia-Carathéodoryschen Satz. Stzgber. Preuss. Acad. Wiss. (1938), 32-48.

Gel'fond, A. O., and A. F. Leont'ef: [1] On a generalization of Fourier series. Math. Sbornik, N. S. **29** (1951), 477-500.

Goodman, A. W.: [1] Univalent functions. Vol. 1; 2. Mariner Publ. Co. Inc., Tampa, Florida (1983), xvii+246 pp.; xii+311 pp.

Gupta, V. P., and P. K. Jain: [1] Certain classes of univalent functions with negative coefficients. Bull. Austral. Math. Soc. **14** (1976), 409-416.

[2] Certain classes of univalent functions with negative coefficients II. Bull. Austral. Math. Soc. **15** (1976), 467-473.

Harnack, A.: [1] Die Grundlagen der Theorie des logarithmischen Potentiales und der eindeutigen Potentialfunktionen in der Ebene. Leipzig (1887).

Hausdorff, F.: [1] Summationsmethoden und Momentfolgen. Math. Z. **9** (1921), 74-109; 280-299.

[2] Momentprobleme für ein endliches Intervall. Math. Z. **16** (1932), 220-244.

Herglotz, A.: [1] Über Potenzreihen mit positivem, reellem Teil im Einheitskreise. Leipziger Ber. **63** (1911), 501-511.

Herzig, A.: [1] Die Winkelderivierte und das Poisson-Stieltjes-Integral. Math. Z. **46** (1940), 129-156.

Holland, F., and D. K. Thomas: [1] On the order of a starlike functions. Trans. Amer. Math. Soc. **158** (1971), 189-201.

Hölder, O.: [1] Über einen Mittelsatz. Göttinger Nachr. (1889), 38-47.

Julia, G.: [1] Extensions nouvelles d'un lemme de Schwarz. Acta Math. **42** (1920), 349-355.

Koebe, P.: [1] Zum Verzerrungssatz der konformen Abbildung. Math. Z. **6** (1920), 311-312.

[2] Über das Schwarzsche Lemma und einige damit zusammenhängende Ungleichheitsbeziehungen der Potentialtheorie und Funktionentheorie. Math. Z. **9** (1921), 77-109; 280-299.

Komatu, Y.: [1] On analytic functions with positive real part in a circle. Kōdai Math. Sem. Rep. **10** (1958), 64-83.

[2] On angular derivative. Kōdai Math. Sem. Rep. 13 (1961), 167-179.

[3] On fractional angular derivative. Kōdai Math. Sem. Rep. 13 (1961), 249-254.

[4] On mean distortion for analytic functions with positive real part in a circle. Nagoya Math. J. **29** (1967), 221-228.

[5] A one-parameter additive family of operators defined on analytic functions regular in the unit disk. Bull. Fac. Sci.-Eng., Chuo Univ. **22** (1979), 1-22.

[6] A one-parameter family of operators defined on analytic functions in a circle. Lecture Notes in Math. No.768, Proc. Anal. Func./Kozubnik 1979,

ed. by J. Ławrynowicz, Springer (1980), 292-300.

[7] Über die Verzerrung des Integrals gebrochener Ordnung der schlichten Funktionen. Bull. Fac. Sci.-Eng., Chuo Univ. **24** (1981), 1-6.

[8] Über die Verzerrung bei konvexer Abbildung des Einheitskreises. Bull. Fac. Sci.-Eng., Chuo Univ. **24** (1981), 7-12.

[9] On length distortion for certain classes of functions analytic in a circle. Bull. Fac. Sci.-Eng., Chuo Univ. **25** (1982), 45-49.

[10] On the range of analytic functions related to Carathéodory class. Ann. Polon. Math. **46** (1985), 145-149.

[11] Über Längen- und Flächenverzerrungen für die Carathéodorysche Klasse. J. Math. Kyoto Univ. **25** (1985), 627-633.

[12] On distortion of the real part in a class of analytic functions related to fractional integration. Compl. Var. **7** (1986), 97-106.

[13] On oscillation in a class of analytic functions related to fractional calculus. Compl. Anal. Appl. '85, Sofia (1986), 327-334.

[14] On a family of integral operators related to fractional calculus. Kodai Math. J. **10** (1987), 20-38.

[15] On the range of values of analytic functions relating to a family of integral operators. Math. Besnik **39** (1987), 399-404.

[16] Note on the product of some integral operators. Saitama Math. J. **6** (1986), 1-7.

[17] On mean distortion with reference to some classes of univalent functions. Unival. Func., Fract. Calc., Appl. Ellis Horwood Ltd., UK (1989), 97-102.

[18] On an analytic differential operator. Unival. Func., Fract. Calc., Appl. Ellis Horwood Ltd., UK (1989), 103-112.

[19] On distortion properties of operators defined on analytic functions. Compl. Anal. and Appl. '87 Sofia (1989), 298-303.

[20] On analytic prolongation of a family of operators. Mathematica, Cluj-Napoca **32** (**55**) (1990), 141-145.

[21] On distortion properties of analytic operators. Kodai Math. J. **15** (1992), 1-10.

[22] On distortion of a class of analytic functions under a family of operators. Proc. of Conf. on Generalizations of Compl. Anal. and Appl. in Phys. (1994), Warsaw/Rynia.

Komatu, Y., K. Niino, and C.-C. Yang (ed.): [1] Analytic function theory of one complex variable. Pitman Res. Notes in Math. Ser. No. 212, Essex (1989), vii+392 pp.

Krzyż, J., and Z. Lewandowski: [1] On the integral of univalent functions. Bull. Acad. Polon. Sci. 11 (1963), 447-448.

Landau, E.: [1] Einige Bemerkungen über schlichte konforme Abbildung. Jber. Deut. Math.-Vereinig. **34** (1926), 239-243.

Landau, E., and G. Valiron: [1] A deduction from Schwarz's lemma. J. London Math. Soc. (4) **4** (1929), 162-163.

Libera, R. J. : [1] Some classes of regular univalent functions. Proc. Amer. Math. Soc. **16** (1065), 755-758.

Lindelöf, E.: [1] Mémoire sur certaines inégalités dans la théorie des fonctions monogènes et sur quelques propriétés nouvelles de ces fonctions dans la voisinage d'un point singulier essentiel. Acta Soc. Sci. Fennicae **35**, Nr. 7 (1908), 1-35.

Livingston, A. E.: [1] On the radius of univalence of certain analytic functions. Proc. Amer. Math. Soc. **17** (1966), 352-357.

Miller, S.: [1] Differential inequalities and Carathéodory functions. Bull. Amer. Math. Soc. **81** (1975), 79-81.

Mogra, M. L.: [1] On a class of starlike functions in the unit disk - I. J. Indian Math. Soc. **40** (1976), 159-161.

Neumann, C.: [1] Vorlesungen über Riemanns Theorie der Abelschen Integrale. 2. Aufl. Leipzig (1884).

Nevanlinna, R.: [1] Über die schlichten Abbildungen des Einheitskreises. Oversikt av Finska Vetens. Soc. Förh. Ser. A **63** (1920), 1-14.

[2] Kriterien über die Randwerte beschränkter Funktionen. Math. Z. **13** (1922), 1-9.

[3] Über beschränkte analytische Funktionen. Commentationes in honorem E. L. Lindelöf, Helsinki (1929), 1-75.

[4] Remarques sur le lemme de Schwarz. C. R. Paris **188** (1929), 1027-1029.

Noshiro, K.: [1] On the theorie of schlicht functions. J. Fac. Sci. Hokkaido Imp. Univ. (1) **2** (1934), 89-101.

[2] On the univalency of certain analytic functions. J. Fac. Sci. Hokkaido Univ. (1) **2** (1934/5), 129-155.

Owa, S.: [1] A note on one-parameter additive family of operators defined on analytic functions. J. Korean Math. Soc. **21** (1984), 177-181.

[2] A remark on the Komatu's conjecture. Math. Japon. **29** (1984), 859-873.

Owa, S., and Zhwo-Ren Wu: [1] A note on certain subclass of analytic functions. Math. Japon. **34** (1989), 413-416.

Ozaki, S.: [1] On the theory of univalent functions. Sci. Rep. Tokyo Bunrika Daigaku (A) **2** (1935), 167-188.

Padmanabhan, K. S.: [1] On certain classes of starlike functions in the unit disc. J. Indian Math. Soc. **32** (1968), 89-103.

Pólya, G. and I. J. Schönberg: [1] Remarks on de la Vallée-Poussin means and convex conformal maps of the circle. Pacific J. Math. **8** (1958), 295-334.

Pommerenke, Chr.: [1] Univalent functions. Studia Math. XXV, Vandenhoeck & Ruprecht, Göttingen (1975), 376 pp.

Robertson, M. S.: [1] On the theory of univalent functions. Ann. Math. **37** (1936), 374-408.

Rogosinski, W.: [1] Über positive harmonische Entwicklungen und typisch reelle Potenzreihen. Math. Z. **35** (1932), 93-121.

[2] Zum Schwarzschen Lemma. Jber. Deut. Math.-Vereinig. **49** (1934), 258-261.

Ruschweyh, St., and T. Sheil-Small: [1] Hadamard products of schlicht functions and the Pólya-Schönberg conjecture. Comm. Math. Helv. **48** (1973), 119-135.

Schur, I.: [1] Über Potenzreihen, die im Innern des Einheitskreises beschränkt sind. Crelles J. **147** (1917), 205-232; **148** (1918), 122-145.

Schwarz, H. A.: [1] Zur Theorie der Abbildung. Programm d. eidg. polyt. Schule in Zürich für d. Schulj. 1869-1870; Ges. Abh. II, 108-132.

Shohat, J. A., and J. D. Tamarkin: [1] The problem of moments. Math. Survey I, Amer. Math. Soc., New York (1943), xiv+140 pp.

Silverman, H.: [1] Univalent functions with negative coefficients. Proc. Amer. Math. Soc. **51** (1975), 109-116.

Špaček, L.: [1] A contribution to the theory of simple functions [in Czech]. Časop. Pest. Math. **62** (1933), 12-19.

Srivastava, H. M., and S. Owa: [1] A certain one-parameter additive family of operators defined on analytic functions. J. Math. Anal. Appl. **118** (1986), 80-87.

Strohhäcker, E.: [1] Beiträge zur Theorie der schlichten Funktionen. Math. Z. **37** (1933), 356-386.

Unkelbach, H.: [1] Über die Randverzerrung bei konformer Abbildung. Math. Z. **43** (1938), 739-742.

Warschawski, S.: [1] On the higher order derivatives at the boundary in conformal mapping. Trans. Amer. Math. Soc. **38** (1935), 310-340.

Weinstein, L.: [1] The Bieberbach conjecture. Intern. Math. Res. Notice **5** (1991), 61-64.

Wolff, J.: [1] Sur une généralisation d'un théorème de Schwarz. C. R.Paris **183** (1926), 500-502.

[2] L'intégrale d'une fonction holomorphe et à partie réelle positive dans un demi-plan est univalente. C. R. Acad. Paris **198** (1934), 1209-1210.

Index of Names

Index of topics —— containing symbols

Other *Mathematics and Its Applications* titles of interest:

C.W. Kilmister (ed.): *Disequilibrium and Self-Organisation.* 1986, 320 pp.
ISBN 90-277-2300-1

A.M. Krall: *Applied Analysis.* 1986, 576 pp.
ISBN 90-277-2328-1 (hb), ISBN 90-277-2342-7 (pb)

J.A. Dubinskij: *Sobolev Spaces of Infinite Order and Differential Equations.* 1986, 164 pp. ISBN 90-277-2147-5

H. Triebel: *Analysis and Mathematical Physics.* 1987, 484 pp.
ISBN 90-277-2077-0

B.A. Kupershmidt: *Elements of Superintegrable Systems. Basic Techniques and Results.* 1987, 206 pp. ISBN 90-277-2434-2

M. Gregus: *Third Order Linear Differential Equations.* 1987, 288 pp.
ISBN 90-277-2193-9

M.S. Birman and M.Z. Solomjak: *Spectral Theory of Self-Adjoint Operators in Hilbert Space.* 1987, 320 pp. ISBN 90-277-2179-3

V.I. Istratescu: *Inner Product Structures. Theory and Applications.* 1987, 912 pp.
ISBN 90-277-2182-3

R. Vich: *Z Transform Theory and Applications.* 1987, 260 pp.
ISBN 90-277-1917-9

N.V. Krylov: *Nonlinear Elliptic and Parabolic Equations of the Second -Order.* 1987, 480 pp. ISBN 90-277-2289-7

W.I. Fushchich and A.G. Nikitin: *Symmetries of Maxwell's Equations.* 1987, 228 pp. ISBN 90-277-2320-6

P.S. Bullen, D.S. Mitrinovic and P.M. Vasic (eds.): *Means and their Inequalities.* 1987, 480 pp. ISBN 90-277-2629-9

V.A. Marchenko: *Nonlinear Equations and Operator Algebras.* 1987, 176 pp.
ISBN 90-277-2654-X

Yu.L. Rodin: *The Riemann Boundary Problem on Riemann Surfaces.* 1988, 216 pp.
ISBN 90-277-2653-1

A. Cuyt (ed.): *Nonlinear Numerical Methods and Rational Approximation.* 1988, 480 pp. ISBN 90-277-2669-8

D. Przeworska-Rolewicz: *Algebraic Analysis.* 1988, 640 pp. ISBN 90-277-2443-1

V.S. Vladimirov, YU.N. Drozzinov and B.I. Zavialov: *Tauberian Theorems for Generalized Functions.* 1988, 312 pp. ISBN 90-277-2383-4

G. Morosanu: *Nonlinear Evolution Equations and Applications.* 1988, 352 pp.
ISBN 90-277-2486-5

Other *Mathematics and Its Applications* titles of interest:

A.F. Filippov: *Differential Equations with Discontinuous Righthand Sides.* 1988, 320 pp. ISBN 90-277-2699-X

A.T. Fomenko: *Integrability and Nonintegrability in Geometry and Mechanics.* 1988, 360 pp. ISBN 90-277-2818-6

G. Adomian: *Nonlinear Stochastic Systems Theory and Applications to Physics.* 1988, 244 pp. ISBN 90-277-2525-X

A. Tesar and Ludovt Fillo: *Transfer Matrix Method.* 1988, 260 pp. ISBN 90-277-2590-X

A. Kaneko: *Introduction to the Theory of Hyperfunctions.* 1989, 472 pp. ISBN 90-277-2837-2

D.S. Mitrinovic, J.E. Pecaric and V. Volenec: *Recent Advances in Geometric Inequalities.* 1989, 734 pp. ISBN 90-277-2565-9

A.W. Leung: *Systems of Nonlinear PDEs: Applications to Biology and Engineering.* 1989, 424 pp. ISBN 0-7923-0138-2

N.E. Hurt: *Phase Retrieval and Zero Crossings: Mathematical Methods in Image Reconstruction.* 1989, 320 pp. ISBN 0-7923-0210-9

V.I. Fabrikant: *Applications of Potential Theory in Mechanics. A Selection of New Results.* 1989, 484 pp. ISBN 0-7923-0173-0

R. Feistel and W. Ebeling: *Evolution of Complex Systems. Selforganization, Entropy and Development.* 1989, 248 pp. ISBN 90-277-2666-3

S.M. Ermakov, V.V. Nekrutkin and A.S. Sipin: *Random Processes for Classical Equations of Mathematical Physics.* 1989, 304 pp. ISBN 0-7923-0036-X

B.A. Plamenevskii: *Algebras of Pseudodifferential Operators.* 1989, 304 pp. ISBN 0-7923-0231-1

N. Bakhvalov and G. Panasenko: *Homogenisation: Averaging Processes in Periodic Media. Mathematical Problems in the Mechanics of Composite Materials.* 1989, 404 pp. ISBN 0-7923-0049-1

A.Ya. Helemskii: *The Homology of Banach and Topological Algebras.* 1989, 356 pp. ISBN 0-7923-0217-6

M. Toda: *Nonlinear Waves and Solitons.* 1989, 386 pp. ISBN 0-7923-0442-X

M.I. Rabinovich and D.I. Trubetskov: *Oscillations and Waves in Linear and Nonlinear Systems.* 1989, 600 pp. ISBN 0-7923-0445-4

A. Crumeyrolle: *Orthogonal and Symplectic Clifford Algebras. Spinor Structures.* 1990, 364 pp. ISBN 0-7923-0541-8

V. Goldshtein and Yu. Reshetnyak: *Quasiconformal Mappings and Sobolev Spaces.* 1990, 392 pp. ISBN 0-7923-0543-4

Other *Mathematics and Its Applications* titles of interest:

I.H. Dimovski: *Convolutional Calculus.* 1990, 208 pp. ISBN 0-7923-0623-6

Y.M. Svirezhev and V.P. Pasekov: *Fundamentals of Mathematical Evolutionary Genetics.* 1990, 384 pp. ISBN 90-277-2772-4

S. Levendorskii: *Asymptotic Distribution of Eigenvalues of Differential Operators.* 1991, 297 pp. ISBN 0-7923-0539-6

V.G. Makhankov: *Soliton Phenomenology.* 1990, 461 pp. ISBN 90-277-2830-5

I. Cioranescu: *Geometry of Banach Spaces, Duality Mappings and Nonlinear Problems.* 1990, 274 pp. ISBN 0-7923-0910-3

B.I. Sendov: *Hausdorff Approximation.* 1990, 384 pp. ISBN 0-7923-0901-4

A.B. Venkov: *Spectral Theory of Automorphic Functions and Its Applications.* 1991, 280 pp. ISBN 0-7923-0487-X

V.I. Arnold: *Singularities of Caustics and Wave Fronts.* 1990, 274 pp. ISBN 0-7923-1038-1

A.A. Pankov: *Bounded and Almost Periodic Solutions of Nonlinear Operator Differential Equations.* 1990, 232 pp. ISBN 0-7923-0585-X

A.S. Davydov: *Solitons in Molecular Systems. Second Edition.* 1991, 428 pp. ISBN 0-7923-1029-2

B.M. Levitan and I.S. Sargsjan: *Sturm-Liouville and Dirac Operators.* 1991, 362 pp. ISBN 0-7923-0992-8

V.I. Gorbachuk and M.L. Gorbachuk: *Boundary Value Problems for Operator Differential Equations.* 1991, 376 pp. ISBN 0-7923-0381-4

Y.S. Samoilenko: *Spectral Theory of Families of Self-Adjoint Operators.* 1991, 309 pp. ISBN 0-7923-0703-8

B.I. Golubov A.V. Efimov and V.A. Scvortsov: *Walsh Series and Transforms.* 1991, 382 pp. ISBN 0-7923-1100-0

V. Laksmikantham, V.M. Matrosov and S. Sivasundaram: *Vector Lyapunov Functions and Stability Analysis of Nonlinear Systems.* 1991, 250 pp. ISBN 0-7923-1152-3

F.A. Berezin and M.A. Shubin: *The Schrödinger Equation.* 1991, 556 pp. ISBN 0-7923-1218-X

D.S. Mitrinovic, J.E. Pecaric and A.M. Fink: *Inequalities Involving Functions and their Integrals and Derivatives.* 1991, 588 pp. ISBN 0-7923-1330-5

Julii A. Dubinskii: *Analytic Pseudo-Differential Operators and their Applications.* 1991, 252 pp. ISBN 0-7923-1296-1

V.I. Fabrikant: *Mixed Boundary Value Problems in Potential Theory and their Applications.* 1991, 452 pp. ISBN 0-7923-1157-4

Other *Mathematics and Its Applications* titles of interest:

A.M. Samoilenko: *Elements of the Mathematical Theory of Multi-Frequency Oscillations.* 1991, 314 pp. ISBN 0-7923-1438-7

Yu.L. Dalecky and S.V. Fomin: *Measures and Differential Equations in Infinite-Dimensional Space.* 1991, 338 pp. ISBN 0-7923-1517-0

W. Mlak: *Hilbert Space and Operator Theory.* 1991, 296 pp. ISBN 0-7923-1042-X

N.Ja. Vilenkin and A.U. Klimyk: *Representation of Lie Groups and Special Functions. Volume 1: Simplest Lie Groups, Special Functions, and Integral Transforms.* 1991, 608 pp. ISBN 0-7923-1466-2

N.Ja. Vilenkin and A.U. Klimyk: *Representation of Lie Groups and Special Functions. Volume 2: Class I Representations, Special Functions, and Integral Transforms.* 1992, 630 pp. ISBN 0-7923-1492-1

N.Ja. Vilenkin and A.U. Klimyk: *Representation of Lie Groups and Special Functions. Volume 3: Classical and Quantum Groups and Special Functions.* 1992, 650 pp. ISBN 0-7923-1493-X
(Set ISBN for Vols. 1, 2 and 3: 0-7923-1494-8)

K. Gopalsamy: *Stability and Oscillations in Delay Differential Equations of Population Dynamics.* 1992, 502 pp. ISBN 0-7923-1594-4

N.M. Korobov: *Exponential Sums and their Applications.* 1992, 210 pp. ISBN 0-7923-1647-9

Chuang-Gan Hu and Chung-Chun Yang: *Vector-Valued Functions and their Applications.* 1991, 172 pp. ISBN 0-7923-1605-3

Z. Szmydt and B. Ziemian: *The Mellin Transformation and Fuchsian Type Partial Differential Equations.* 1992, 224 pp. ISBN 0-7923-1683-5

L.I. Ronkin: *Functions of Completely Regular Growth.* 1992, 394 pp. ISBN 0-7923-1677-0

R. Delanghe, F. Sommen and V. Soucek: *Clifford Algebra and Spinor-valued Functions. A Function Theory of the Dirac Operator.* 1992, 486 pp. ISBN 0-7923-0229-X

A. Tempelman: *Ergodic Theorems for Group Actions.* 1992, 400 pp. ISBN 0-7923-1717-3

D. Bainov and P. Simenov: *Integral Inequalities and Applications.* 1992, 426 pp. ISBN 0-7923-1714-9

I. Imai: *Applied Hyperfunction Theory.* 1992, 460 pp. ISBN 0-7923-1507-3

Yu.I. Neimark and P.S. Landa: *Stochastic and Chaotic Oscillations.* 1992, 502 pp. ISBN 0-7923-1530-8

H.M. Srivastava and R.G. Buschman: *Theory and Applications of Convolution Integral Equations.* 1992, 240 pp. ISBN 0-7923-1891-9

Other *Mathematics and Its Applications* titles of interest:

A. van der Burgh and J. Simonis (eds.): *Topics in Engineering Mathematics.* 1992, 266 pp. ISBN 0-7923-2005-3

F. Neuman: *Global Properties of Linear Ordinary Differential Equations.* 1992, 320 pp. ISBN 0-7923-1269-4

A. Dvurecenskij: *Gleason's Theorem and its Applications.* 1992, 334 pp. ISBN 0-7923-1990-7

D.S. Mitrinovic, J.E. Pecaric and A.M. Fink: *Classical and New Inequalities in Analysis.* 1992, 740 pp. ISBN 0-7923-2064-6

H.M. Hapaev: *Averaging in Stability Theory.* 1992, 280 pp. ISBN 0-7923-1581-2

S. Gindinkin and L.R. Volevich: *The Method of Newton's Polyhedron in the Theory of PDE's.* 1992, 276 pp. ISBN 0-7923-2037-9

Yu.A. Mitropolsky, A.M. Samoilenko and D.I. Martinyuk: *Systems of Evolution Equations with Periodic and Quasiperiodic Coefficients.* 1992, 280 pp. ISBN 0-7923-2054-9

I.T. Kiguradze and T.A. Chanturia: *Asymptotic Properties of Solutions of Nonautonomous Ordinary Differential Equations.* 1992, 332 pp. ISBN 0-7923-2059-X

V.L. Kocic and G. Ladas: *Global Behavior of Nonlinear Difference Equations of Higher Order with Applications.* 1993, 228 pp. ISBN 0-7923-2286-X

S. Levendorskii: *Degenerate Elliptic Equations.* 1993, 445 pp. ISBN 0-7923-2305-X

D. Mitrinovic and J.D. Kečkić: *The Cauchy Method of Residues, Volume 2.* Theory and Applications. 1993, 202 pp. ISBN 0-7923-2311-8

R.P. Agarwal and P.J.Y Wong: *Error Inequalities in Polynomial Interpolation and Their Applications.* 1993, 376 pp. ISBN 0-7923-2337-8

A.G. Butkovskiy and L.M. Pustyl'nikov (eds.): *Characteristics of Distributed-Parameter Systems.* 1993, 386 pp. ISBN 0-7923-2499-4

B. Sternin and V. Shatalov: *Differential Equations on Complex Manifolds.* 1994, 504 pp. ISBN 0-7923-2710-1

S.B. Yakubovich and Y.F. Luchko: *The Hypergeometric Approach to Integral Transforms and Convolutions.* 1994, 324 pp. ISBN 0-7923-2856-6

C. Gu, X. Ding and C.-C. Yang: *Partial Differential Equations in China.* 1994, 181 pp. ISBN 0-7923-2857-4

V.G. Kravchenko and G.S. Litvinchuk: *Introduction to the Theory of Singular Integral Operators with Shift.* 1994, 288 pp. ISBN 0-7923-2864-7

A. Cuyt (ed.): *Nonlinear Numerical Methods and Rational Approximation II.* 1994, 446 pp. ISBN 0-7923-2967-8

Other *Mathematics and Its Applications* titles of interest:

G. Gaeta: *Nonlinear Symmetries and Nonlinear Equations.* 1994, 258 pp. ISBN 0-7923-3048-X

V.A. Vassiliev: *Ramified Integrals, Singularities and Lacunas.* 1995, 289 pp. ISBN 0-7923-3193-1

N.Ja. Vilenkin and A.U. Klimyk: *Representation of Lie Groups and Special Functions.* Recent Advances. 1995, 497 pp. ISBN 0-7923-3210-5

Yu. A. Mitropolsky and A.K. Lopatin: *Nonlinear Mechanics, Groups and Symmetry.* 1995, 388 pp. ISBN 0-7923-3339-X

R.P. Agarwal and P.Y.H. Pang: *Opial Inequalities with Applications in Differential and Difference Equations.* 1995, 393 pp. ISBN 0-7923-3365-9

A.G. Kusraev and S.S. Kutateladze: *Subdifferentials: Theory and Applications.* 1995, 408 pp. ISBN 0-7923-3389-6

M. Cheng, D.-G. Deng, S. Gong and C.-C. Yang (eds.): *Harmonic Analysis in China.* 1995, 318 pp. ISBN 0-7923-3566-X

M.S. Livšic, N. Kravitsky, A.S. Markus and V. Vinnikov: *Theory of Commuting Nonselfadjoint Operators.* 1995, 314 pp. ISBN 0-7923-3588-0

A.I. Stepanets: *Classification and Approximation of Periodic Functions.* 1995, 360 pp. ISBN 0-7923-3603-8

C.-G. Ambrozie and F.-H. Vasilescu: *Banach Space Complexes.* 1995, 205 pp. ISBN 0-7923-3630-5

E. Pap: *Null-Additive Set Functions.* 1995, 312 pp. ISBN 0-7923-3658-5

C.J. Colbourn and E.S. Mahmoodian (eds.): *Combinatorics Advances.* 1995, 338 pp. ISBN 0-7923-3574-0

V.G. Danilov, V.P. Maslov and K.A. Volosov: *Mathematical Modelling of Heat and Mass Transfer Processes.* 1995, 330 pp. ISBN 0-7923-3789-1

A. Laurinčikas: *Limit Theorems for the Riemann Zeta-Function.* 1996, 312 pp. ISBN 0-7923-3824-3

A. Kuzhel: *Characteristic Functions and Models of Nonself-Adjoint Operators.* 1996, 283 pp. ISBN 0-7923-3879-0

G.A. Leonov, I.M. Burkin and A.I. Shepeljavyi: *Frequency Methods in Oscillation Theory.* 1996, 415 pp. ISBN 0-7923-3896-0

B. Li, S. Wang, S. Yan and C.-C. Yang (eds.): *Functional Analysis in China.* 1996, 390 pp. ISBN 0-7923-3880-4

P.S. Landa: *Nonlinear Oscillations and Waves in Dynamical Systems.* 1996, 554 pp. ISBN 0-7923-3931-2